From entrepreneurs and venture capitalists, to artists, social critics, and researchers on the cutting edge of technology . . .

When entrepreneurs and electronics moved to sleepy Santa Clara County, California, and begat Silicon Valley, the world would never be the same again. This is where the most brilliant and unconventional minds of our time have created machines that will transform our lives, and a hard-driving, free-wheeling culture that may revolutionize our future.

What is it like to live in a society where everyone eats, sleeps, and breathes technology—where bakers offer "silicon chip" cookies, and streets have names like "Disk Drive?" Mahon, an industry insider, gives us candid, in-depth profiles of industry leaders, whiz-kids, and longtime observers that reveal the men and women who manipulate electrons and money as you have never seen them before.

These are the citizens on the edge of the world—a group of people as dazzling and diverse as the products they make and market. This book is their portrait: a searching, perceptive, and often surprising record of a community where everyone lives for the future.

CHARGED BODIES

THOMAS MAHON is a writer and public relations consultant whose clients include several Silicon Valley firms. He and his wife live with their two sons in the San Francisco Bay area.

Technology Today and Tomorrow from MENTOR and SIGNET

CHARGED BODIES

People, Power, and Paradox in Silicon Valley

Thomas Mahon

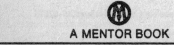

A MENTOR BOOK

NEW AMERICAN LIBRARY

NEW YORK AND SCARBOROUGH, ONTARIO

MENTOR TRADEMARK REG. U.S. PAT. OFF. AND FOREIGN COUNTRIES
REGISTERED TRADEMARK—MARCA REGISTRADA
HECHO EN CHICAGO, U.S.A.

SIGNET, SIGNET CLASSIC, MENTOR, PLUME, MERIDIAN AND NAL Books are published *in the United States* by New American Library,
1633 Broadway, New York, New York 10019,
in Canada by The New American Library of Canada Limited,
81 Mack Avenue, Scarborough, Ontario M1L 1M8

Library of Congress Catalog Card Number: 85-62614

First Mentor Printing, April, 1986

1   2   3   4   5   6   7   8   9

PRINTED IN THE UNITED STATES OF AMERICA

To my family

There is no history, only biography.

—attributed to Ralph Waldo Emerson

| CONTENTS

CHARGED
BODIES

1 | WHAT'S GOING ON HERE?

1 | Overview

It's dangerous to pull off to the side on the winding roads in the Santa Cruz Mountains at night. You run the risk of being sideswiped by another motorist who comes barreling around a curve. But for the view, it's worth the risk.

I find a place to pull over, end my ascent and get out of the car. A steep, forested mountainside lies to the south and the west. Behind it and above it is a black and starlit sky. This could be the most deserted place in Arizona, Utah or Wyoming. It's only when I walk across the road and face to the north and east that the brilliant and sprawling valley two thousand feet below abruptly comes into view: the San Francisco Bay Area.

A stunning sweep of orange light runs from Leland Stanford's old farm in Palo Alto on my far left, fifteen miles or so south by southeast into San Jose on my right. Behind the arc of light is blackness again: the Bay and the expansive tracts of salt evaporators. Rising beyond that is a scattering of lights on the lower slopes of the Diablo Range in the East Bay.

Even at this late hour, I sense a gigantic engine idling. It seems I can hear the dynamo bristling in place, waiting to come fully to life again when the sun comes over the Diablos in a few hours.

A place should have a name before it has a nickname.

Officially, this is called the Santa Clara Valley. It stretches between the Santa Cruz Mountains and the San Francisco Bay, from Palo Alto, through Mountain View, Sunnyvale and Santa Clara, down to San Jose. Then continues on, beyond what I can see, about as far south again toward the farming community of Gilroy.

The northern half of the valley has another, more contemporary identity. These two hundred or so square miles of orange street lights down below me are better known as Silicon Valley.

That half-county down there saw the development of the integrated circuit, the microprocessor, the personal computer and the video game, not to mention such esoterica as distributed data processing, the Winchester disk drive and the plug-compatible mainframe.

Silicon Valley is symbolic of more than the latest developments in electronic and computational wizardry. It marks as well an economic and cultural frontier where successful enterpreneurship and venture capitalism, innovative work rules and open management styles provide the background for what is perhaps the most profound and far-reaching inquiry ever into the nature of intelligence. Combined with research here in bioengineering and "artificially intelligent" software, this inquiry may extend dramatically our sense of who we are and what we can do, and affect our very evolution.

If the changes promised by Silicon Valley were only of a technical or commercial nature, this place could be compared to Birmingham or Manchester, Pittsburgh or Detroit. But a visitor who spends some time here—exploring the labs and

Overview

executive suites, the bars and garages—would come away with the sense that a time, a place and a group of people have come together as in Periclean Athens, Renaissance Florence or postwar Paris to affect world culture.

For good or ill? That judgment can't yet be made. And by the time it can be made, the products and technologies generated in, and represented by, Silicon Valley will be all-pervasive. This technology controls the beating of arrhythmic hearts; brings air travelers safely though crowded skies; stands to revolutionize the concept of classroom, workplace and entertainment; cooks meals while the diners are at work; credits and debits bank accounts without any exchange of greenbacks or silver; and makes long-range weather forecasts that affect food prices, foreign trade and geopolitical strategy.

The same silicon-based semiconductors that make all that possible, however, also guide nuclear missiles and allow bombs to become "smart."

These individual applications are merging and becoming all-encompassing, until electronic devices stand to occupy the center of our culture, affecting art, aesthetics, philosophy and perhaps theology, as much as our medicine, transportation and defense. And that development—that heritage of Silicon Valley—may prove to be every bit as profound as the creation of tools or the invention of writing.

Parallel with the possible misuse of the technology, as a matter of concern, is the ever-widening gap between the advancement of that technology and the limited comprehension of it by people it affects.

It's as if our world were coming to run simultaneously on two clocks, or live by two calendars. The technocrats are on one time; everyone else is on another. Dialogue becomes increasingly difficult. As difficult as it would have been for a Victorian to explain his age's advances in psychology, engineering and the social sciences to one from the Middle Ages. Not only would it be hard to explain the specifics, but the very concept of such progress would have been difficult to convey to one from the static medieval world.

Silicon Valley represents a midpoint in the progression between King Arthur's plaintive plea for magic—"Merlin, make me a hawk so I can fly away!"—and the calm request of the starship's Captain Kirk—"Beam me up, Scotty."

Who are these technocrats and what is this place? What is its role in the evident transformation of our world, and how did it come to assume that position? Surely the full responsibility for the electronic and computer revolution can't all be attributed to these few square miles.

Indeed, neither the first electronic components nor the first computer systems came from here. And today, major semiconductor foundries are located in Texas and Arizona, while established manufacturers of mainframes and minicomputers are headquartered in New York, Pennsylvania, Minnesota, Massachusetts and Japan. The trade press that covers the electronics and computer industries is, for the most part, centered in New York and Boston.

What then is Silicon Valley's claim to fame? It's the place where the man who co-invented the semiconductor chose to start his own company; it's half a county which, having virtually no industrial base of its own as recently as World War II, saw nearly seventy other semiconductor businesses spin into existence from that first company between 1955 and 1980; it shares the honor of developing an advanced type of semiconductor that integrates many electronic circuits onto a single piece of silicon; then it saw the development of the microprocessor that, in effect, integrates a number of integrated circuits onto a fleck of pure sand.

Like that protozoan that belongs in both the plant and animal kingdoms, the microprocessor belongs in the domains of both electronic components and computer systems. And because that "computer on a chip" came from Silicon Valley, the microcomputer industry—the mass manufacturing and mer-

Overview

chandising of personal computers—grew up around here, too.

Those computers are powerless without programs to direct them. So a microcomputer software industry developed here as well, creating programs for business, for learning, and for fun and games.

Work is also being done to create software that is "artificially intelligent," that allows computers to be self-correcting: able not only to compute, but to compute about their computing; possibly able to create their own programs.

Software exists on a medium—either "floppy" or rigid disks, or silicon chips. A number of companies were started here to make both the memory media and the hardware to access the stored information.

Then, too, the microprocessor has found other applications besides driving computers, in such fields as telecommunications, computer-aided design, industrial control and robotics. Companies in these industries are here, as well. And all of these companies—whether hardware or software vendors—have their support services: for machining, calibrating, tooling, documenting, publicizing, accounting, litigating, registering.

The microprocessor allows genetic engineers to handle the massive amounts of information they deal with. Many of the key patents in this new field are held by institutions in this area. So Silicon Valley promises to become Siliclone Valley, too. In fact, even now, thought is being given to merging recombinant DNA research with advances in semiconductor design to create "biochips," or molecular electronics devices. Combine that with robotics and artificial-intelligence software, and we may someday see "living computers." Should that ever happen, how will we define ourselves thereafter?

Clearly, the impact of Silicon Valley, and what it represents, goes far beyond new technology development and new enterprise formation, to hint at some major rethinking of our self-perception.

*   *   *

What's often missed in all the microprocessor-based, user-friendly, ergonomically designed, polysyllabic jabberwock of high technology, however, is the lubricating effect that blood, sweat and tears have on the free flow of electrons.

The human drives in this community are as diverse as the disciplines of the new technologies themselves. People here are driven by greed and ambition, by hunger and desperation, by a quest for style and elegance and celebrity status. Not unlike anywhere else. But here, unlike elsewhere, the end result of all this enterprise is the height of impersonality and anonymity: "black boxes" with poetic sobriquets like the 2100 Z/A. The tension between the impersonality of the end products and the passion to create them in this single-focus community is fascinating.

As diverse as they all are, however, the new technologies ultimately trace their roots back to semiconductors, those tiny devices made of pure sand from which this valley takes its name, and on which its reputation originally rested. This place is, literally, a kingdom built on sand. As our world is, increasingly, a society pervaded by and built upon those tiny chips of inorganic silicon that appear to mimic the thought processes of our minds.

How does a semiconductor come to pervade our lives? Consider that a spring is a memory device. Stretched or compressed, it returns to its original shape. But it takes a lot of space to accomplish that, and it has some inherent inefficiencies. Consider replacing that spring—in an automobile's suspension, for example—with a sophisticated memory chip. In one more area, then, our dependence shifts from something we can see and understand to a device whose operation most people take on faith. And until such time as there is at least some general understanding of the basics of electronics by laypeople, the mechanical and the comprehensible will continue to be replaced by the seemingly magical. Ironically, the ultralogical world of electronics may lead us to a new age of superstition.

Overview

The community of Silicon Valley is the first in the world in which virtually the whole society—the economy, the idle chatter, the world outlook—revolves around the new electronic technologies. There are lawyers here, but they help new technology companies get started; the local politicians go to Washington to urge incentives for entrepreneurs; bakers offer not chocolate chip, but silicon chip cookies, boxed like floppy disks; much of the medical research here is microprocessor-based; philosophers speculate on the locus where the studies of high-energy physics and theology intersect; such literature as comes from here is destined for the pages of industrial and trade, not literary, magazines; composers create with computers; and the great academic institutions here have, by and large, made their reputations in their engineering and business schools. This is the place where "high tech" is making its first appearance as one of the liberal arts.

In some respects, Silicon Valley is a state of mind. Other places, trying to recreate the phenomenon, hope to attract technical industries with claims of being "Silicon Desert" (Arizona), or "Silicon Forest" (Oregon), or "Silicon Glen" (Scotland). But something unique and unprecedented has happened here, the impact of which goes far beyond this valley. The world of the late twentieth and early twenty-first centuries will certainly owe its ultimate shape to the technologies that first flowered here.

What's to become of this valley, and of us all everywhere, as a result of the work and the attitudes generated there? Does this place have any connection to the larger America beyond these bordering mountains—the America of the Grand Canyon and the Blue Ridge Mountains, the Rust Belt and the South Bronx? Or did it just spring spontaneously into being sometime between 1955 and 1975, the offspring of some gigantic computer-aided design mechanism, independent of any past?

No, that is definitely not the case. For as I look below me, one of the brightest of those night-lit streets is the spine of this valley's development, its link to a long-vanished past. El Camino Real was first trod by eighteenth-century Spanish Franciscans in the service of God and king, as they founded the missions around which grew such communities as San Jose, Santa Clara, San Carlos, San Mateo, San Bruno, San Francisco.

How ever did Mission Santa Clara metamorphose into today's industrial suburbs? That story begins even before the padres, with pacific natives living in Edenlike innocence; and continues after the military, cultural and religious domination of the Spanish, to an idyllic, arcadian Californio society based on horsemanship and hospitality; to the later Yankee discovery near here of man's ultimate fantasy, a mountain of gold; to the dream-racked sleep of a bereaved railroad baron; to the frustrated boosterism of local promoters; to tenacious young men at the turn of the century transfixed by the seemingly miraculous phenomenon of wireless communications; to an engineering professor who took the base notion of applied technology and graced it with all the academic legitimacy in the world. From this valley, and those Berkeley hills across the Bay, came two seminal technologies, radar and atom smashing, that helped win World War II for the Allies. All of this before anyone ever heard of such a thing as a transistor.

Between the roots of this place and the products of its labs and assembly lines there has developed here a community in some ways like, and in some significant ways unlike, any other community. Ever. This book is a portrait of that community *as* a community; a self-portrait, actually, drawn by those who shape and share this place.

What's my interest in all this? I came here, from much farther east, some years back to manage a public relations agency that represents companies in this valley. I'm a layman; I didn't have a technical background. But what struck me was that my work was giving me an opportunity to witness the center of a revolution—cultural as much as technical—from the inside.

Overview

I am an outside agent called in to live for a while at the center of a company and help shape its identity as a technical, business and social entity. But because I'm an outside agent who moves on, and because of the pace at which things happen here, there is little time to retreat from the day-to-day, to pull back and get a larger view such as this nighttime mountainside provides. That's what compels me to undertake this project now. To pull back from the details of my daily work, writing about the intricacies of LAN and CAD/CAM and VLSI and CMOS, and instead find out, from the denizens and habitués of this most remarkable place, what it's all about, in more detail than I see from the freeway and in less detail than I care to gather from spec sheets and tech manuals.

This is not a book about chips. Rather, it's about a community of people who care passionately about technologies and a society built upon chips.

2 | Paradox Valley

As an outsider without a technical background who has come to see some of this place from the inside, there is only one word I can think of to describe a stranger's initial reaction to Silicon Valley: paradox.

Begin with the basics. Silicon Valley is currently making over our world with electronics. And the upshot of it is, atomic particles, of which the electron is one, represent true enigmas. For these dazzling, dancing charged bodies exhibit properties of both wave and particle, and thereby pose a superb conundrum.

Measured with one type of instrument, they demonstrate one set of properties; measured with another instrument, the same particles exhibit opposite properties. Clearly our understanding of the world at the subatomic level is incomplete. But for the present, we are left with this paradox: atomic particles like the electron now seem to refute classical logic, which has held that a thing can't be itself and its opposite—can't simultaneously exhibit opposite properties. Yet today we use the electron to make logic devices function logically.

So much for conundrums at the micro level; they exist on a macro scale, too. It has been proposed that all technology—whether high or low—exists to make life more predictable. Yet, paradoxically, in Silicon Valley the nature of the technology businesses is the ultimate in unpredictability, with fortunes made and lost in less than two years.

The paradoxes lead to other paradoxes. The community of Silicon Valley prides itself on being "open" in the way Californians have of being open: congenial, relaxed and laid back. The companies here even try to institutionalize that openness, offering their employees flextime, company-sponsored beer busts, an insistence on first names all around, and such democratic customs as unpretentious cubicles and unreserved parking slips even for top executives.

Yet Silicon Valley is not really all that welcoming to the uninvited newcomer. A midday drive through the industrial parks in this place is an eerie experience. There are few sidewalks here. There are even fewer pedestrians. Except for the cars on the roads, there is almost no sign of life at ten o'clock on a workday morning in the center of this dynamic, gigantic industrial engine.

The buildings are pleasant enough, built on a human scale, unlike Manhattan's towers. But like the urban skyscrapers, they reveal absolutely nothing of what's inside.

Showing the Spanish influence on local architecture, the offices and factories turn their backs on the streets and instead turn in upon themselves, often facing an inner courtyard.

Except for the parking lots overflowing to the street, a visitor would never know anything was going on behind the inpenetrable walls. Should he go to the front entrance, and ask to go inside to be assured there really are people there, the polite but firm guard will politely but firmly send him back to the corporate property line. And that will put him back in the street. (There are few sidewalks, remember.)

In this place where trade secrets are the secret of success, it's possible to drive mile after mile and never once see a welcome sign.

Paradoxically, this industrial inhospitality is found in one of the most inviting places on earth to live. The weather in the Santa Clara Valley represents one of the purest forms anywhere of a climate type called Mediterranean, considered the finest our bodies can enjoy.

Latitudinally, the area exists between the great rain forests of the Pacific Northwest and the dry heat of Baja and Sonora. From east to west, too, the environment is moderate. The Coast Range, or the Santa Cruz Mountains, on the San Francisco Peninsula creates the western wall of the valley, protecting it from the winter fury of the Pacific and the ocean's summer fog, while the Diablo or Mt. Hamilton Range to the east keeps both the winter tule fog and the summer heat locked in the inland valleys.

For three months of the year, the prevailing westerlies bring the rains that turn the golden hills to Irish green. Then for the remaining nine months, the Pacific High-Pressure System keeps the climate temperate, the air cool and dry, and the skies clear. (Although not as clear as in the past, since local wind currents make the South Bay a catch basin for the auto and industrial emissions of the entire Bay Area.)

The local vegetation is a rich profusion of tropical, subtropical and boreal: redwood, oak, olive, palm and wine grape. And though you wouldn't know it for all the parking lots, the soil here is some of the most fertile in the world. The early American settlers were delighted to find the area supported an abundance of fruit trees, such as apricot, peach, plum, prune, cherry and pear.

The appealing, congenial nature of this valley is reflected even in the place-names: Sunnyvale, Mountain View, and the memory of such saintly patrons as gentle Clare of Assisi, follower of St. Francis, and mild Joseph, father-protector of the Christ child.

Paradox Valley

And not only is this place alluring in itself, it is surrounded by an incredible array of attractions: Big Sur and Carmel; Mendocino and the Napa Valley; Yosemite and Tahoe.

Paradox builds upon paradox. Few communities in history have witnessed such economic success in so short a period of time. In the past five years, over a thousand new technology-oriented businesses have started here.

Yet few places ever suffered such a long spell of frustrated ambition, waiting for its chip to come in, through three civilizations and nearly two centuries. The Spanish looked in vain here for an earthly paradise, then had their empire carved up by the Mexicans; who, in turn, saw their land grants stripped away by the intruding Yankees; whose merchants, in turn, begged without success for over a century to have outside industry bring jobs into the area.

So the star of this technology-focused community is clearly on the rise. But how long can you build an economy based on information processing when the country is cutting back on producing the things the information represents, like cars and ingots of steel and pairs of shoes? Can you really sustain, over an extended period, an industrial community whose stock in trade is the accumulation, processing, storage and retrieval of information designed to accumulate, process, store and retrieve information?

Again paradoxes. Here is a community affecting the culture of the world, from personal computers to video games to the components that drive every consumer electronic product bringing art to the outback. Yet this place is, by virtually any accepted definition, a cultural wasteland. By and large, the fine arts don't exist here. Nor are they missed.

A common passion is to increase net worth, in communities named after Saints Clare and Francis, who turned away from all material gain.

*   *   *

Paradox Valley

There is one last paradox that conveys the enigma at the root of the Santa Clara/Silicon Valley. Long before the Spanish arrived here, this area was settled by a tribe of native Americans called the Costanoans, or, sometimes, the Ohlone. It's thought they came here from Asia, over the Aleutian land bridge, as many as five thousand years ago.

Over the span of more than a hundred generations, their burial mounds show absolutely no progression beyond bows and arrows, mortars and pestles. The Spanish, when they came, thought them the most primitive of all the peoples they encountered in the new world. The climate here was so congenial, game and water so plentiful, belligerency, stress and necessity so unknown to them, that the Costanoans never developed the concept of change or progress or invention.

The captain of the first Spanish ship into the San Francisco Bay remarked that these natives were "constant in their friendship and gentle in their manners." And because they were so accommodating, so pacific, so gentle, they had their way of life stripped from them by the conquering Spanish.

Accustomed to going naked, they were made to wear clothes. Used to worshiping outdoors, they were converted in churches and made to live in enclosed mission compounds. If they escaped to the hills that had once been their home, they were rounded up and brought back in chains by soldiers with nothing better to do.

They saw their land renamed by the Spanish, from the San Francisco Bay, to Santa Clara, to the Santa Cruz Mountains, to Monterey Bay. Between the diseases brought by the Spanish, for which they had no natural immunity, and the later Yankee concept of good Indians being dead Indians, the Costanoans were doomed. Within a hundred years of first meeting Europeans, between the late eighteenth and late nineteenth centuries, they became virtually extinct. One has to search long and hard in Santa Clara County today to find a clue to the existence of a people who lived here for perhaps as long as five millennia.

Paradox Valley

Yet while those natives never heard terms like "state of the art" or "leading edge," and never got beyond the mortar and the pestle over scores of generations, they understood, as today's Valley residents do, the concept of living at the brink. The Costanoans found satisfaction in stasis, as the technologists here today find it in dynamism. Technology sometimes threatens to be our undoing; lack of it was theirs. Both dwell on the edge.

The Costanoans knew the worth of their lives, the significance of what they did. And they sensed the magnificence of their location in the natural world. Before their tragic encounter with the Europeans, these children of Asia would gather on the shores of the Monterey Bay, turn their backs to North America and face the sundown sea. And they would begin to dance a long dance, and as they did, they would chant over and over what could well be the chant of the radio engineers and microelectronics engineers, the nuclear physicists and the genetic engineers, the exobiologists and the artificial intelligencers, and all the others who are part of the modern story of this place: "We are dancing, we are dancing on the brink of the world."

2 | THE ELECTRON MANIPULATORS AND DEAL MAKERS

3 | Entrepreneurship

What drives Silicon Valley at the micro level is the electron, that marvelous enigma that fires the material world and defies our logic. What drives Silicon Valley at the macro level is "the deal": the business plan that describes a product, defines a market, and charts the growth course of a proposed new company.

Electron manipulating and deal making have been going on here since just after the turn of the century, but never so much as in recent years. From Hyatt Rickeys in Palo Alto to Maxi's at the Red Lion Inn in San Jose, the deals propel advances in manipulating the electron, and further familiarity with the electron opens up new territory for making deals.

Sometimes the deal maker sees a need for a nonexisting product and scouts for an electron manipulator to design it. Sometimes the latter—the technical wizard—goes in search of the former to support an idea for an electronic component, a computer system, a software package. And on occasion, both roles are performed by the same individual—the deal-making electron manipulator. However it's done, that symbiotic combination of the technical and the financial, that promotion of the enterprising by the venturesome, is the foremost activity of Silicon Valley.

How prevalent is all this? In peak times, at least five new technology companies a week start up here. And while this phenomenon is certainly going on elsewhere, in no other place is the entire culture so centered around, so attuned to, so reinforcing of the electron-manipulating/deal-making entrepreneurial activity. Silicon Valley is a one-industry town. And the growth industry here is growing new industries.

What is it that the enterprising engineers want? Why do they "go for it" with such a vengeance? To become rich? Certainly that's part of it. What's surprising, however, is what a relatively small fraction of the drive is devoted simply to increasing one's net worth.

Besides, if you want to stand out as rich in Silicon Valley, you have to be *really* RICH. I was having lunch with a man who was president of a very successful new company. To my chagrin, I discovered that my credit card was over my limit. My lunch companion, wanting to spare me any embarrassment, told me the same thing had recently happened to him, except—he said without pausing for effect—his personal credit limit is $50,000. And, wouldn't you know, he thought he had the full line of credit, having just paid his bill, but it turned out his wife had recently bought something that put him over the top.

What, I wondered, can one's wife buy for $50,000 that one wouldn't be aware of?

Several years before, this entrepreneur quit a salaried job, helped found a computer company, and became worth over $10 million when that company sold its stock to the public. In his early forties, he quit that company to risk half of his new fortune to start another enterprise. When asked why not retire

Entrepreneurship

comfortably for the rest of his life, he looks at me uncompre-hendingly. That option isn't even an option to him.

Becoming a millionaire a dozen times over in less than fifty months is not that remarkable an accomplishment here, as my friend pointed out. When his TeleVideo Systems be-came a publicly held corporation, founder Philip Hwang's stock was reported in *Fortune* magazine to be worth $794.6 million.[1] Now, *that's* rich. But if the drive here were only for that sort of financial success, few but the most deluded would even set off to rival it. And if there's one thing entrepreneurs aren't, it's deluded. Irrational, cantankerous, petulant and headstrong sometimes, but never deluded.

For all the surface glitter, and all the talk about money here, it's not the acquisition of wealth that seems to be foremost in the minds of the entrepreneurs. It is, rather, an even more fundamental drive: to create. To make a new type of semiconductor, a new model of computer system, a new software program. The drive to doodle and tinker and make a technical breakthrough, however, is only the visible manifes-tation of an even deeper type of creativity: the need to make order out of chaos; to make as broad a reality as possible out of hope and pluck and determination; to create a microculture that promotes creativity in others. It's done for two reasons.

Twenty years ago, engineering students were usually identi-fiable from a hundred yards away. They wore thick glasses, lugged oversized briefcases, unself-consciously displayed slide rules, and tended not to be the ones taking out the homecom-ing queen.

Silicon Valley is where, by various and diverse routes, many of those people eventually gravitated. They came to California, got contact lenses, trimmed down, toned up, got tan, and set out to create not just new products, but a new environment.

Consciously or unconsciously, they set out to create uncorporate corporations, loosely structured structures, to do away with the pecking order they had known "back there"

—wherever "there" was—where jocks and kids with family names were big shots, and where others were consigned to the periphery.

The second reason for creating a relatively freeform culture here is that to survive in the technology businesses, a company has to create breakthroughs on a routine basis. And you simply don't promote that kind of original thinking in a traditional orders-from-on-high environment. Technology companies *have* to be creative, simply for the privilege of staying in business to be able to create again.

Even at the largest, most mature Silicon Valley companies, the formalities and privileges of business organizations elsewhere are generally absent. Company founders in their seventies, like David Packard and William Hewlett, still prefer to be called by their first names, even by new-hires. Intel's cofounder and vice-chairman, Robert Noyce, occupies an open workspace with only shoulder-high partitions separating him from a hundred similar work stations all around it. An on-site, $1.5 million recreation center at ROLM is available for all employees (and families on weekends), and is not reserved strictly for director level and above. The one executive dining room I am aware of in Silicon Valley exists in a company that is owned and controlled by a larger firm in the East.

There are exceptions to the egalitarianism, to be sure, but generally the pomp and circumstance of other business cultures, whose management styles embrace a few and frustrate many, is absent here.

The people most responsible for all of this entrepreneurial activity today—these enterprising electron manipulators who make new technical devices as well as new social environments—combine single-mindedness with a mastery of myriad skills and a good deal of intuitive finesse.

To know how to spot a window of opportunity before it opens and be ready to jump through it at the first crack; to know when to leave the security of an existing job and when

Entrepreneurship

to stop refining a design and take it into production; when to enter a market; when to turn over some authority to an administrator; how to price a product for a market that doesn't exist yet; how to hire and keep and motivate good people; how to make a worldwide corporate family of ten thousand feel as close as the original corporate family of five; how to gauge when you have no more to contribute and how to find your own replacement; how to manage in the ebb and flow of fast-moving markets and the ups and downs of a global economy and, after all that, never lose sight of what was once no more than a private fantasy . . . to succeed at doing all of that well is every bit as rewarding, stimulating, compelling as any drive to acquire several millions of dollars.

Probably the most important skill a start-up team needs is resource orchestration. Consider a few of the areas which an entrepreneur has to understand to take a private vision and make it into a public reality: manufacturing resource planning; employee benefits; facilities expansion; marketing strategies; order entry systems; foreign distribution; motivation; assimilation; negotiation; judicious procrastination. Chief executives of large, mature corporations have teams of seasoned experts to deal with each of these areas. Often, if the top executive of a Fortune 500 company gets involved at all, he or she only approves a decision made by specialist subordinates.

The entrepreneur, on the other hand, may head an organization of only three to eight staffers, exhausted from working sixteen-hour days, when he or she is called upon to make decisions in each of these areas, often in rapid-fire succession. And with each decision, the entrepreneur plays "bet the company." Any one wrong decision could bring the whole thing down . . . if not immediately, then, like a time bomb, later when there is even more at stake. Who can be conversant in all these fields, each of which is, after all, a side issue to the new company's main business?

In one week's time, I saw two company founders here deal with problems that represent the range of issues they confront.

One had to get on the phone himself to yell and scream at the president of a firm that makes shipping crates, lambasting him for the fact that the imprint of his new company's logo on the crates looked nothing like the logo that was supplied as a model.

The next day, I met a man who was delighted that his component business was growing so fast he needed to open another manufacturing facility, but disappointed over the fact that the plant was to be in Wales. "It really sticks in my craw that I'm creating jobs for Welshmen when there's forty percent unemployment in Detroit. But the Welsh government makes it so attractive to go there, with subsidies and tax advantages. . . . In comparison, our government penalizes me for staying here, creating jobs for Americans."

The entrepreneur confronts so many areas beyond his control, yet he has the ultimate responsibility for each decision. And the most sobering thought of all is that by the time someone sets off to start a new enterprise, it's too late to learn new skills. With the pace and pressure that comes with the job, there is only time to capitalize on skills already possessed.

Entrepreneurship is not unique to Silicon Valley, but it is more common here today than anywhere on earth. Elsewhere, it's still the exception; here it's a general obsession.

My job has given me the chance to work with, and watch at close hand, the workings of companies here at every stage of their development. At one end of the spectrum is the individual, demonstrating a hand-wired prototype of a new modem on his kitchen table, hoping to persuade me to represent him in exchange for now valueless shares of stock. As we talk, and he meticulously prepares espresso for both of us, he waits . . . and waits . . . and waits . . . for a potential backer to return his call.

Midway along the start-up curve is the company president who is oblivious to the stripped-down "tilt-up" production facility he rents in an office park in Milpitas. Rather, he acts

Entrepreneurship

like the chief executive of the $100 million disk drive company he expects this to become within three years. He no longer wears the cotton slacks from the days when his office was his home, but rather a suit of sincerest blue.

Finally, at the far end of the spectrum is the president/founder of a minicomputer company—twelve years beyond the start-up phase—who says with great relish that he doesn't have an entrepreneurial bone left in his body. And then proceeds to chortle over the fact that when he was a supplicant at the venture capital trough a dozen years before, he vastly overstated his capital requirements so that he would not only get more than he needed, but also deplete the pool of money available for possible competitors.

I am called into these and other situations to help articulate the persona of the entrepreneur/president and his enterprise at that stage in their evolution.

The first man needed some—any—publicity for his new product in the electronics trade press. Since he couldn't seem to attract venture capital backing, he hoped that a timely new-product announcement would generate enough sales to launch the company on its own, and let him "bootstrap" its growth on sales alone.

The second president, with secured venture capital backing and recently moved into new facilities, needed coverage in the local papers to attract good employees in a sellers' job market. He also wanted to convey the perception of mature solvency to the business and trade press, so that large-volume original-equipment manufacturers who are his potential customers would be assured they were dealing with a solid organization (only seven months old).

The third president, after watching a twenty-month downturn in his company's fortunes, needed "to push iron out the door," to increase sales immediately. He had to regain the respect of the financial community before the next annual meeting, eight weeks hence, at which time, as things stood, he fully expected shareholders to demand his resignation.

I find I've developed a split reaction to these people. One that parallels the paradox of so much passion and fever and fury being invested in products that are so cool and analytical and emotionless.

Professionally, I have to be detached from the company's situation enough so that however grandiose the entrepreneur's self-perception becomes, I must, of all the court retainers, play the Fool and say, "Look, don't kid yourself, this is what you can reasonably expect others to say about you."

But however removed from the situation my work compels me to remain, it's hard on a personal level not to get caught up in the risk, the rush, the anxiety and the euphoria of watching people create a product, maybe a whole new industry, not to mention a social/business environment, out of whole cloth.

In our world of big institutions and faceless bureaucracies, it would be hard for anyone not to empathize with an individual or a small group that has tossed security over, and burned every bridge, to pursue a plan in the face of overwhelming odds and obstacles. In a way, these enterprising engineers resemble classical heroes. They take the drives that everyone is subject to and act them out on a larger-than-life scale, where virtues seem more virtuous, vices more vicious.

Take humility. I had lunch with the founder/chief executive officer of one of the Valley's bigger firms, then doing about $200 million a year in sales. We arrived twenty minutes late for our reservations at the restaurant, and as we waited for the maître d' to seat us, he confided nervously, "I hope they won't be mad at us for being late. I hope we haven't inconvenienced them." Out the window of the seafood restaurant, as a backdrop to his anxiety, was the six-building, twenty-acre headquarters complex of the company he'd started.

If that kind of self-effacement makes one feel warm and fuzzy about company starters, it should be pointed out they can also act as tyrants. There is one president who had this signed message taped to the mirror in every washroom in

Entrepreneurship

what he considered his corporate fiefdom: "BY MY DECREE, you are only to use one paper towel when you wash your hands." And he meant it. And he was obeyed.

Probably the single characteristic that classical heroes and entrepreneurs have most in common is the ability to risk it all when so much is beyond the individual's control. The entrepreneur who took his chances and did everything right, with the best of motives, can watch his world dissolve under him within six months. In other, more stable industries, that businessman and his offspring might have known two or three generations of solvency. In the technology industries, things move so fast it can all go away in less than a year. And if he doesn't actually go out of business, the entrepreneur may be relegated to "the land of the living dead," that no-man's-land where he is able to keep his doors open, but is never again thought of as being in the first tier, or able to attract the brightest people or the biggest customers. Fortune smiles on computer people here with the same fickleness she once reserved for rock stars.

The president of a very successful microcomputer company threatened to sue a client of mine, the founder of a new microcomputer company, because the new company's name was close to that of the established company. The threat of that lawsuit made my client pale. A year later, the established company had set up a creditors committee to pay off its debts, had laid off over sixteen hundred people, and was facing a class-action suit itself. The start-up, meanwhile, was still in business and just shipping its first products.

These comments on entrepreneurship are from an observer. Participants, of course, have much more insight. Roy Dudley is a self-described "start-up mechanic." He has started three Silicon Valley firms, grown them to a certain level, then sold them and moved on. For Dudley, the rush isn't to build an organization and cash in big at the end. Rather, it's to lay the foundation for someone else to build upon. About the time he's expected to begin wearing a tie every day, Dudley moves on, taking a "healthy chunk of change" for his start-up efforts. His first San Francisco Peninsula company built a powerful portable water purifier and was sold to a larger firm in Wisconsin. Then he and a partner started another business to manufacture telecommunications devices. That company, too, was sold to an existing firm named Plantronics. He's now involved in a third start-up, and in a later chapter will recount his experiences in more detail.

For now, by way of introduction to self-profiles of Silicon Valley company founders, this is a "professional entrepreneur" describing his work and his world; an environment that has all the elements of a Tolkien fantasy, of a boom-town heyday, and of a gangster movie, once you enter what Dudley calls "the entrepreneurial swamp."

"It's that invisible place where you go away from the normal, workaday world; where you trade security for hope. That's the swamp. You tell the rest of the world that you see a different path. You say, 'Look, I'm going to go out and work on something that has no prior value. I'm going to burn up a year or two of my life. I'm going to put fifteen or twenty thousand of my own cash into this, and take a pass on making forty or sixty grand at Hewlett-Packard, with the distinct possibility that I'm going to get down to the other end and have it just be a big wash. Boy, am I going to look dumb!

"That's the entrepreneurial swamp. You crawl in not knowing where or when or *if* you're ever going to crawl out. But the catch is that you're going to meet a lot of fascinating creatures along the way. Because once you do leave the workaday world, it turns out that there are a whole lot of people out here that are part of this twilight zone. There are guys that make their living being part of this. And there's a lot of money that flies around out here."

Dudley is unabashedly romantic about his work. To him, there's a mystique about it all, playing the role of the high-

Entrepreneurship

tech bandito, the last American cowboy, possessing "the Silicon Valley Easy Rider mentality." He finds it fun to be able to be out and about, walking around the Valley, while so many others are deskbound. But at the same time, knowing that he is, in fact, doing something productive: creating a framework, establishing the networks, making the connections necessary to form something out of nothing. He compares this business environment here to San Francisco in the gold rush, and Hollywood in the early '20s.

"There's a quickening of the pulse, of people's willingness to want to become part of it. But you should know some things before you get into this game. There's a classic line. Someone in the Hell's Angels got picked up for taking this thirteen-year-old girl up to Oregon. He got caught on the Mann Act and he was facing charges of contributing to the deliquency of a minor. He decided to be his own defense and he got up in front of the court and said, 'If she didn't want to go fast, she shouldn'ta oughta got on.' I think that's true in this business, too."

Dudley's advice to someone in a secure job is to think long and hard before ditching it all to climb into "the swamp."

"Consider what's at stake. If you've got an idea for a new kind of widget, you'd better do your homework. Because once you get out there, you clip off that security and you go for the hope. It just might all screw itself right into the ground."

The entrepreneurs of Silicon Valley have far more in common with artists than with the inhabitants of the corridors of corporate power. Artists and entrepreneurs inhabit a twilight zone. They live in daily chaos, pushed by doubts, drives, exuberance, obsessions, and tension, to create a new reality out of an idea, talk and hope. Life in the swamp and life in the studio causes sleepless nights, pleas for understanding, the ability to string out creditors, make believers out of skeptics, and beg and borrow and skam and jive to keep the sumbitch alive, just one more day.

The artist and the entrepreneur each see before them a fixed lodestar. For the best of them, the vision never varies, which is what allows them to pursue it with such single-mindedness. It's the ones who change the dream daily who fail to build anything on shifting sand.

Entrepreneurial opportunities come and go. Twenty years ago, it was starting semiconductor foundries. Soon the established companies raised the ante, and the admission price got too high. So ten years ago, those with entrepreneurial drive went on to making small computer systems, based on the ready availability of those semiconductors. Then that market, too, got flooded with well over 150 vendors. After that the opportunity was microcomputer software, which runs on the microcomputer systems built upon the components. It's a logical progression.

What's next? Entrepreneurs without engineering or programming backgrounds have already begun to step in to adapt computer power to specific fields: medicine, agriculture, service businesses. Entrepreneurship will probably cease to be so closely tied to a technology orientation, but will rather consist in implementing technology elsewhere.

Three fates await the determined entrepreneur. Going out the bottom: failing to get the vision off the ground. The side: seeing his talents can take him no further and having to turn the enterprise over to a more able administrator. And the top: selling the organization and moving on to start another, or taking a comfortable retirement.

And all along the way, three abiding fears nip at the entrepreneur's heels.

First, there's the danger, as Apple found, of being too successful in showing that there is a huge market for a heretofore nonexistent product like the personal computer. Suddenly, a Big Blue Giant named IBM rouses from its

Entrepreneurship

slumber, decides that market rightfully belongs to it, and uses its virtually limitless resources to try to take over the market in eighteen months. To match wits with IBM, or other large organizations, the entrepreneur has to play poker with an opponent who has the resources to say, "I'll see your dime, and raise you ten thousand dollars." Anyone in those circumstances, however ballsy otherwise, finds it a chastening experience: "Geez, is this royal flush really good enough to hold with?"

The only thing that saves entrepreneurs at all is that in the time it takes large corporations to shift their enormous mass and change course, the entrepreneur can bob and weave and dodge and feint and move on before he's stepped on. Free to go off and try it again.

Of course, there's the distinct possibility that companies like IBM let them get away, to sweat out the trail-blazing phase at their own expense and prove there's a business there, in order to pave the way for their considered Grand Entrance.

Then there's burnout, the sudden or gradual shift from gut-twisting tension to nerve-deadened listlessness, when nothing seems worth anything anymore. The loss of appetite, the boredom, the entropy that descends upon someone whose system finally rebels against a steady diet of sixteen-hour days, every day, and 3,000 psi of pressure per minute.

For those who burn out at the end of the course, there's early retirement or buy-out. But for those who burn out along the way, without the accomplishment yet in hand, there's an element of tragedy. This man or woman challenged the status quo with everything in his or her life. And fell short.

The entrepreneur who successfully navigates all the pitfalls— creates a product; begins a business; makes the transition from start-up to stability; avoids crippling lawsuits, burnout and technical obsolescence—is frequently confronted—and betrayed—by his own example. A key person within the organization becomes similarly afflicted with entrepreneurial enthusiasm. At the very least, the person will take his needed skills and leave. Possibly with company plans, secrets and other key employees. Sometimes to become a direct competitor.

The only consolation that Entrepreneur Number One has is that Entrepreneur Number Two will someday confront Entrepreneur Number Three. Roger Borovoy, former general counsel at Intel, related the rule of the road for entrepreneurs in Silicon Valley: "Don't let your employees do to you what you did to your former boss."[2]

Roy Dudley's assessment of entrepreneurial drive is more gritty:

"Silicon Valley is littered with the bones of people that went out and tried and failed. You know, I've been through some real hard times. I've been bankrupt. But something kept me from going back and getting a job. It's just something about having to go and work at a routine job that is unacceptable for me.

"I had a friend who was in the rackets and he said, 'If you can't do the time, don't do the crime.' And maybe that's it. Maybe there is a certain amount of punishment that goes with starting companies."

4 | Frank Deverse; Huey Lee

Frank Deverse is one of dozens of "Fairchildren," former employees of Fairchild Semiconductor of Mountain View, who, since 1959, have gone off on their own to start and build other semiconductor companies in the Silicon Valley. His International Microcircuits Incorporated, or IMI, designs and manufactures a type of semiconductor called a gate array.

Deverse is not as well known or as flamboyant as some of his Valley colleagues. And the fact that he is neither of these makes him representative of the typical Silicon Valley entrepreneur. The unprepossessing nature of most of them stands in direct contrast to the spectacle of their collective accomplishments.

He is one of the relatively few who is both electron manipulator and deal maker, having been his company's sole founder and growing the business without the help of outside investors. He's an engineer by background who got his formal business education on the job, as president of his own company. And while that company is successful now, Deverse is familiar with business failure.

He is, like most of the people here, a transplanted Easterner who has come to love this part of the country. He's a congenial man in his mid-forties, with bearded Italian features, and a strong independent streak.

In IMI's offices, which occupy a cluster of three buildings in a Santa Clara business park, he talked about his migration westward from Pennsylvania, and upward from corporate researcher to company president.

"I grew up in western Pennsylvania, was a maverick, refused to go to college. My father begged me, told me he would buy me a new car, but I still refused to go to school because I just didn't have confidence in myself. Got married and then started to wonder how in the hell was I going to support my wife. I didn't even have a job. Decided I had better go to school, after all."

He went to Duquesne, majoring in chemistry, but it was a stressful experience, with the course load, a full-time job, and a young family of two children, each one born during finals.

"I go to school. It's on a mountain. I have to walk up there and lug these thirty pounds of books and I can't breathe by the time I get to the top. I am twenty-four, supposed to be in good health. I go to the doctor. He says, 'You've got the worst case of nerves I've ever seen. Take these pills and you'll feel better.' I took the pills and sure enough, I felt better, except I didn't give a damn about anything. I just didn't want to study. I said, 'Well, it's either be nervous and finish this school or flunk out.' Threw away the pills and went through."

After college, he secured a research job at Gulf Oil, but found that in a facility full of Ph,D.s, having a B.S. degree was little better than being a technician. It was while on that job, in 1963, that he read a book on transistors. "I was fascinated that you could engineer molecules, atoms. I said,

Frank Deverse; Huey Lee

'Hey, that's really an exciting business.' So I found a job in the business.''

He moved to IBM and went to work running a portion of a pilot semiconductor manufacturing line, building new parts. Within four years, he secured eighteen patents and tripled his salary.

"It was all the fun I anticipated. I really enjoyed the hell out of it. I made my own decisions. They just happened to be different from other people's. I got disciplined many times. For instance, I made a decision, which I shouldn't have, to put something I developed on my own on the pilot line. I converted the whole damn line over to doing it. And it was running away, doing terrific, for about six weeks, and one day the manager finds out. He calls me in and says, 'What in the hell are you doing?' So I told him [about a new process to keep moisture off the surface of silicon]. He said, 'How's it working?' I said, 'It looks like it's working terrific.' He said, 'We ought to patent it.' ''

And that's just what his manager did, adding his own name to the patent application, and splitting with Deverse half the $50,000 from IBM. That process is now used throughout the semiconductor industry.

In the late 1960s, a group of sixty-six men left IBM to start another company in Wappingers Falls, New York, called Cogar, with the aim of making low-cost memory semiconductors. Cogar was started at the same time as Intel in Santa Clara, to do exactly the same thing but using a different process.

"We felt we were going to win. We built a superior product to Intel's. But we made all the wrong business decisions. We priced the product wrong. We solicited the wrong customers. We offered them the wrong things. We just did everything wrong. We went out of business. Twenty-seven million dollars in three years. The company went bankrupt.'' The incident had a traumatizing effect on Deverse.

"I wanted to stay in the semiconductor industry, and

Frank Deverse; Huey Lee

Fairchild offered me a job. I came out here and joined
Fairchild Semiconductor, and found it to be very exasperating,
undisciplined. My manager told me, 'Okay, we are going to
work Saturdays and Sundays. We're going to run through
walls. We're going to pull things off out of brute force.' I had
just come out of Cogar, where I'd heard too many of those
stories, and I said to myself, very quietly, 'Bullshit. He might
run through that wall, but I ain't running through that wall.'

"Well, I wasn't ready to leave yet, so I made it look like I
was running. Then I would go back to my office and close
my door and say, 'They are all crazy, and I am not going to
do that.'

He decided after his second day he was going to leave, but
at the time he didn't know where to go next. He started IMI
forty-five days after joining Fairchild Semiconductor, incorpo-
rating under his lawyer's name, running an answering service
and going home to return the calls every lunchtime, then
going back to work in the afternoon.

Six months later, in February 1973, he quit Fairchild to
devote full time to IMI. The company started out making
something called a hard surface photomask, to be used in the
manufacture of integrated circuits. But because he saw that as
a contracting market, Deverse decided to begin making integrated circuits themselves. However, since he lacked the
resources to compete against the big, established semiconductor houses, he decided to address a specialized niche of the
market: small-volume users who need customized or semi-
customized circuits.

In the early '70s, if a customer wanted only ten thousand
custom-made semiconductors, he would have to buy a hundred thousand units and throw nine-tenths of them away. It
would take that large an order to interest a supplier. There
was no way then for the small-volume, quick-turnaround
customer to get satisfied.

"We thought customized circuits would be a good business.

Frank Deverse; Huey Lee

We had no product, but we ran an ad and we got five hundred phone calls in two weeks. We said, 'Oops, there's really a market out there.' I guess it's unethical to advertise a product that we didn't have, just to see what would happen. But we got so much response, we decided to go through had build it.''

The product IMI offered, called a gate array or logic array, is a partially completed semiconductor whose final configuration is set to a specific customer's request. Today it is one of the fastest-growing segments of the semiconductor business, and is expected to be a $5-billion-a-year industry by 1990. When Deverse started, his was the only independent company making the product.

''I was scared to death when I left Fairchild, because I hadn't really gotten over the failure of Cogar. It left an emotional scar. One thing I learned at Cogar, I was never going to take a job strictly for money. I took that job to succeed, and when it didn't succeed, I decided I would never take another job just to get rich. I would have to enjoy it. If I made money, wonderful. But in the end if I didn't make money, I was still going to walk away and say I enjoyed it.''

The pressure that had plagued Deverse as a student, that afflicted him with Cogar's failure, came back to haunt the entrepreneur. This time, a negative cash flow—more money going out than coming in—was the problem. ''I made the decision and discussed it with my wife. 'If we don't get over this hump, I am going to back out. I can't take this pressure anymore.' ''

Just in time, in 1978, IBM introduced its 3300 computer system, which made use of gate-array technology. When IBM, in effect, granted its imprimatur on the technology, the entire world of semiconductor users took note. IMI's prospects changed overnight. And a new industry was born. ''After six years, we finally got over the hump and we became cash-solvent. We were able to have a working margin, which took a lot of pressure off me.''

Frank Deverse; Huey Lee

IMI was the first digital gate array manufacturer. Today there are about one hundred competitors that have come into being since 1978. IMI is still one of the top three.

"The only way I was able to pull it off was that I got into the market before '78, when the company was developing very slowly from its own earnings without outside investors. In fact, it was developing as fast as I had cash to respond to it. I was learning how to do business. By the time the market exploded, I had about graduated from business school."

From his own experience, and watching the growth of other companies in his business, Deverse has identified several stages a growing company goes through. First there is the Existence Stage, where creating a product and identifying a market for that product are the chief concerns. The Survival Stage begins when there are some customers and a minor return on the initial investment, but continued existence is still far from assured. The Success Stage begins when the company has sufficient size, market acceptance and profits that quarter-to-quarter existence is no longer a worry. Now a decision must be made before entering Stage Four that will determine the long-term fortune of the enterprise.

The entrepreneur can be exhausted and choose to retire; can determine the company will do well to hold its place; or can pull together all his cash and borrowing power and "take off." In opting for the third course, as he was doing at the time we met, Deverse sees the key problems of the Takeoff Phase as being orchestration of resources, particularly people.

"What I've said to my top nine guys is, 'Fellows, there is a Fifth Stage which I call Resource Maturity. This is where we want to go, and this is really big business. And you are either going to become part of making the decisions and making this company go forward to that end, or I am going to have to replace you. Some of you have come out of jobs where you never had this opportunity, you were the executor. Well, here you're at the upper level of management and you are expected to generate ideas. If you can't, then I've got to assume you have overstepped the limits of your talents. Either step aside and take a job where you can succeed . . . or you are going to have to leave.' ''

Addressing his managers this way was not an easy thing for the normally affable Deverse to do.

"I'd been rolling this around in my head for six months. How to do this in a civilized, professional, efficient, humanistic manner, instead of having each one of these potential dismissals be emotionally destructive to this person. In fact, I didn't do it until I got comfortable with it.''

The rolling of heads is not a fate reserved only for managers in rapidly growing start-up companies. Sometimes the founder-turned-president can't make the transition from collegial entrepreneurship to corporate executive. Some understand that themselves and bow out gracefully. In some cases, the outside investors fire the founder.

Deverse, who avoided venture capital because he didn't want that outside control, sees that the time has come for him to begin divesting his authority to a team of managers who can handle the responsibility of running the larger organization that he is determined IMI should become.

"Now I have to begin letting go of this business I started, and I have to trust these guys to make decisions. Some won't make it. I am going to judge them based on whether they succeed, but they should not take it so damn personal.

"I wish somebody had told me all this at Cogar. I would have walked out of there without kicking myself because I would have understood it there. I think firing is usually done in terms like 'You failed,' or 'You're dumb,' or 'You're incompetent.' What it should be is, 'Hey, we tried you at a higher level and your talents didn't match this job. We gave you a shot.' If you look at it that way, it's more palatable. You can swallow that. What it comes down to is, 'I reached higher than my talents. Nothing wrong with reaching. Aspira-

Frank Deverse; Huey Lee

tion is terrific.' But be prepared that you aren't always going to succeed.

"I have a few people to whom I am giving an opportunity to operate at a very high level. Some won't be capable of operating there. They just don't have the imagination, the drive, the intelligence they aspire to. So I've said, 'Look, if your talents aren't optimum, but if you want to make up for it with hard work, sixty, seventy hours a week, that's okay. That's your business if it fits your life-style. If your wife puts up with it. If your children will tolerate it.' "

The latter course is not one that Deverse views with a lot of optimism. The overworker, who tries to make up for a lack of creative ability with a constant dose of long hours, is leaving himself open to the pain of burnout: the continued sense of frustration that leads to numbing ineffectiveness.

"I feel absolutely no inkling of being burned out as long as I am moving forward, making progress and making decisions that are timely, appropriate and show constructive results. If any of those things don't happen, then I become frustrated and I want to get away from that pain."

Burnout is not, of course, unique to Silicon Valley, but it seems to be more common here because there are so many rapidly growing companies and the demand for top-rank managers far exceeds the supply. In any population, there are only so many excellent people to go around. And so less skilled or less mature people are pushed into responsible positions before they are ready for them, or have had the chance to grow into them.

Mature companies can afford to nurture their managers; start-ups haven't the time for that luxury and instead put people on the firing line with a firm belief in on-the-job training. That approach doesn't always work out for the best.

"Some of the managers then take off and go to Idaho to run a country store, which I think is wrong. They should have stepped back and been more realistic about what their talents really were. I know what I'm really good at and I know what

Frank Deverse; Huey Lee

I'm poor at. I try to do what I'm good at. I try to aim my job and my life at my strengths, not my weaknesses.''

Orchestrating resources—particularly people: one's self and one's employees—is probably the single biggest skill an entrepreneur needs, particularly in the technology businesses, where managing creative human resources counts for so much more than in the traditional industries that are concerned primarily with processing natural resources.

"I would never have been able to achieve this company if I'd stayed in Pennsylvania. This place contributed to my learning how to run a business in a very short period of time. The collective ideas that circulate in this valley contribute to the people who are trying to do something. It happened in my case. I see it happen to other people."

Fine for Silicon Valley, but can this industrial bonanza work elsewhere?

"I think you can set the stage for people to take a chance. You make support facilities easy to come by. Maybe Pennsylvania should pick metal processing or foundry work . . . what they're really good at. You can get molds made there, you can get designs made, you can get all this made and put it in this collective area. And then say, 'We're going to emphasize this and now anybody who wants to take a crack, go do it.' That's how it happened here. Like a nuclear reaction, it starts to feed on its own energy after a while."

Why does Deverse think he adapted so well to this California setting?

"I think this valley is a concentration of people who want to build things. I tried to build shacks when I was a kid. I must have built twenty of them. I spent my whole childhood building, making things, constructing things. To tear something down eats away at me. I feel a real gnawing inside. It busts something. It breaks me up when a glass breaks. I think that must be encompassed in my drive to build the business. I have just got to believe that is part of it. I haven't figured it all out yet, but I know that's there. I feel it more and more as I get older.''

If orchestrating limited resources is a key to success in Silicon Valley—or in any emerging enterprise, for that matter—then the creation of an appropriate corporate culture has to be of equal importance.

Huey Lee has attempted to do just that at his Advanced Technical Services, which is located in Fremont, a community across the Bay from overcrowded and high-priced Santa Clara. It is in East Bay towns like Fremont and Milpitas that the second wave of Silicon Valley start-up activity is currently taking place.

Lee began his engineering career in the aerospace industry, but the secrecy involved in working with defense electronics, plus the desire to work instead with life-saving devices, prompted him to switch to work in the design and marketing of medical electronics products for a leading firm in that field. At the same time, seeing he would eventually like to start his own business, he began to study for his master's degree in business administration at the University of Santa Clara.

He continued to work at his job by day, and spent his nights, until three or four in the morning, designing and building his own product to calibrate EKG monitors in hospitals.

Eventually, the pressure and the exhaustion of his split existence were too much for him.

''I decided that wasn't good enough. I couldn't support my family working only part-time on my new business. I decided that if I were to do any justice to my own plan, I'd have to quit my job. So in February 1977, I plunged in with two feet, with a savings of about ten thousand dollars, and started this company.

''My wife, May, came into the business to help me. Hocked my house, everything. I was doing the sales and engineering while she was running production. We grew every year. The

Frank Deverse; Huey Lee

first year, it was eighty thousand dollars. The second year, a half million dollars. The third year, a million and a half. Now we're about four million. And employ about three hundred people.''

Advanced Technical Services is now set up to do electronics assembly on a subcontractor basis for the electronics companies in the Bay Area—stuffing chips into printed circuit boards according to the customer's scheme. It was a good time for Lee to get into that business in the late 1970s. Chastened by the recent recession and the resulting layoffs, many electronics firms in Silicon Valley decided that henceforth they would only staff up to a certain manpower level, then vend out to a firm like ATS any electronics assembly work over and above that level. That way, in the event of another business downturn, the manufacturing companies would not have to face laying people off. That becomes Huey Lee's problem.

A number of companies have started up in recent years to do electronics assembly work on a contract basis. In some cases, they have become the scandal of the electronics industry. The work of stuffing chips into boards, in intricate patterns in rapid succession, is menial, mind-numbingly repetitive, and in some cases done in working conditions that are downright Dickensian. Undereducated or overeducated people, speaking no English, are consigned to dead-end jobs in some pretty grim working conditions. The effects on workmanship are telling; the effects on lives are telling as well.

Lee's facility, on the other hand, is clean, well lit and spacious. His employees—Mexican, Chinese, Vietnamese, Filipino—attend to their tasks with such detail that returns for defective workmanship are far lower than industry averages. The effects of the environment are evident not only in the quality of the work, but in the fact that employees bring other members of their families to the company, as well. The story of how Lee motivates his employees to be so meticulous doing such repetitious work, how he has created a positive culture in an otherwise drab business, can be traced back to his native China.

Lee's father grew up on a farm in Sacramento and graduated in accounting from the University of California at Berkeley in 1926. Being Oriental, however, he was not able to find a job in his field then and had to take work as a butcher in a Chinese grocery. "He got fed up with that because he didn't go to school for four years for that. However, there weren't any other jobs for him . . . because of discrimination. So he tore up his citizenship paper and went back to China with a hundred dollars in his pocket. Quite an adventurer."

Because of his American education, the senior Lee had a succession of good jobs, including comptroller of the Chinese Air Force, and eventually the general manager of a bank. He was with that bank when the communists came to power.

"In 1948, we fled to Hong Kong, where we were refugees for seven years as noncitizen, nonstatus persons. We went to school in Hong Kong, and he was working for the same bank there as a manager. When we all graduated from high school, he had no money to put us through college there because it was very expensive. So he decided to apply for refugee status and come back to America so that we could continue our education. He sacrificed his good job and came back here for us. But he was too old to get a job, so he borrowed money from relatives and started a small mom-and-pop grocery in Oakland. We all helped in that store, and it supported us through our school years. Finally, when all of us kids graduated from college, we closed it down and we supported our parents."

Lee and his wife decided when they started their company that they would treat their employees as they had wished to be treated in their own previous jobs. This evolved into a company philosophy which they call Theory C—where the C stands for Common Sense.

"There's no magic about it. You try to understand how

people feel and try to make them feel happy when they are working. I have a dentist here now, a business owner, a professor from Vietnam, a school principal from China. People from all walks of life are coming to me. The only drawback is they can't speak English, but we have interpreters. And so we help them get a job, a good job, and at the same time, learn English.''

His employees, all legal immigrants, are often vastly over-qualified for the work. But they can't take up their old professions in this country until they are certified. And they can't begin that process until they learn the language and the customs. In the meantime, they have to support themselves and their families.

''We provide them with a beginning. It isn't easy. The work here is very repetitive, but in every repetitive job you can find challenge. You can find it in how to do the job better, how to be faster, how to improve quality and find new assembly techniques. We shift people around to learn every process.''

Lee's attempt to create a decent, humane working place extends in many directions. ATS subsidizes over 50 percent of the employees' hot-lunch program. There is subsidized gasoline for workers. There are employee-of-the-month, perfect-attendance and zero-defects awards, and Safeway gift certificates; and, as much as possible, promotions are from within. When outside managers are hired, they have to begin by working on the assembly line themselves for three months. ''So they can understand how it feels, what working conditions are, so they can manage better in the future. This program is very successful. The managers are very, very sensitive to the employees' needs.''

Huey Lee is a gentle, soft-spoken, firm man. The rapport he has with his employees seems to be based on genuine interest. He introduced me to his employees at a weekly coffee-and-doughnut gathering: first in English, then Spanish, then Chinese.

As with most Americans today, all of my family's records from the early years in this country are gone. I can only imagine about someone named Mahon or McMahon, first name unknown, coming through a place like Ellis Island into what must have been a terrifying new world. I was mindful of that, standing in a company cafeteria full of Mexicans and Orientals. These men and women—many of them extremely overqualified—are doing what they have to do so that a generation or two from now names like Lee and Garcia and Vong will seem as familiarly American as Murphy and Minetti are now.

Out the cafeteria window, across the Bay in Santa Clara, fourth-generation Americans hustle deals and manipulate electrons. These men and women in Fremont are paying the entrance admission that the entrepreneurs' great-grandparents paid long ago. But the entrepreneurs and the immigrants have vastly more in common than might first meet the eye. Both groups have taken their lives, and the destinies of their descendants, into their hands and said, ''It's going to be different after this.'' Those who only know the mainstream might find that a cliché. For those who are living it out today, the transformation has rather more immediacy.

Having lived in both East and West, Lee has some ideas of what global cultural considerations are involved with entre-preneurship.

''I went to high school in a different culture. I could not choose my subjects. I was told, 'This is what you have to study.' We went to school six days a week. Here, high school students have the liberty to choose subjects. This is why I think the high schools fail here. When my daughter went to junior high, she brought home a list of classes that she wanted to take and I took one look and they were all easy subjects. No math, no science. I asked her, 'Why aren't you taking any of these other subjects?' She said, 'Nobody takes them. They're too hard.' I said, 'You are going to take math and science whether you like it or not.' Now she is thanking me."

Yet the very thing for which Lee faults American high schools, he applauds in this country's colleges.

"I feel this is the best college system that there is. People who go to college *want* to go to college. They are not forced to go to college. You can proceed at your own pace. You are not forced by peer pressure, or by face-saving or whatever. This is not the case in Hong Kong, where you are forced to take subjects. It impedes your creativity. This is why in America there are so many creative people. I think the elective system is good at the college level when people are eighteen or older."

Lee has other thoughts on the nature of entrepreneurship. "I'm a Catholic. It's very hard to be an entrepreneur and a good Christian. Not to say that entrepreneurs are not ethical. It is that Christian and Buddhist philosophy taught people to be passive and not to be venturesome. And an entrepreneur is basically a venturesome person. Ethical or unethical is beside the point here. A true Buddhist or a true Christian tends to be passive. They are not aggressive.

"My feeling is this. Stanford, Harvard train proper managers. They don't train entrepreneurs. Schools such as San Jose State, where I went to college, are not well-known, and they tend to produce more entrepreneurs. The reason is that the students there know that they couldn't compete with Harvard's name or with Stanford's name. In order for them to get from Point A to Point B without that help, they have to do something themselves.

"You take a Stanford graduate and you can always find him at the height of the corporation. And you take a guy from San Jose State who is an average student to start with, and comes from a school that is not well known, and who tries to

Frank Deverse; Huey Lee

do something himself . . . he runs into a lot of obstacles in the corporate world. There is no chance that I can be president of a big corporation. Let's face it . . . I have no background, no history, I didn't come from a wealthy family. In order for me to get there, I've got to do it myself. I always wanted to give it a try. If I don't succeed, I can always fall back on my capabilities. Confidence is what I have.''

5 | Roy Dudley

Roy Dudley, the pathfinder through the entrepreneurial swamp whom we met before, matriculated into the
world of high-technology data and telecommunications devices via the auto industry and the low-tech field of saltwater
purification.

Because Dudley is a start-up mechanic—only taking his
companies to a certain level before selling them—he is atypi-
cal of most Silicon Valley entrepreneurs, for whom the challenge is to grow the business to Fortune 500 class.

But because Dudley has been through the entrepreneurial
swamp on several occasions, he's seen what general qualities
are needed, and which are to be avoided, for success. He
describes those findings in a voluble, gregarious, highly charged
way that one would expect of a man whose profession is
"rolling the bones."

He's probably as close as they come to being a "pure
play" in entrepreneurship. He is an engineer, though not an
electronics engineer, as well as a deal maker. His special

Roy Dudley

forte is resource orchestrating, particularly in the freeform, early days of the company when one has to be adept at everything. His jumping-off point, however, coincides with the establishment of a regularized corporate culture.

When we met, Dudley was between start-ups. It was a sunny, summer Saturday morning, and from his Daly City home we could look out on the Pacific and the distant Marin Headlands.

"My first experience with having to go from a concept to a finished product in a finite amount of time came in 1969 in my senior year in college. We had to do a senior project, make some application out of what we'd learned in the first three years. There were a couple of us that were inherently lazy and we thought, 'Let's put a project together that's so outrageous that there's no chance we could finish it. They will have to grade us on a partial completion and we'll get our degree.' "

Dudley and his colleagues petitioned the National Science Foundation for a research grant to build an electric car. Surely, they thought, just processing the paperwork would take all year. Time would be up and they would be able to collect their degrees at year-end without having to do any project. To their utter astonishment, the group received a grant of $10,000 within two weeks. Further, they were told they had exactly four months to build and drive a fully electric automobile and enter it in the Great National Clean Air Car Race, between MIT in Cambridge and Cal Tech in Pasadena.

"There was a philosopher who said once that the prospect of going to the gallows the next morning tends to focus the mind wonderfully. Our degrees were on the line! It was my first experience with 'Don't let your mouth write a check that your ass can't cash.' "

This was Dudley's first entry into the entrepreneurial mind-set, the world of "whatever it takes," where there are no

Roy Dudley

rules. All he knew was he had to be in Boston in four months with a car that was fully electric.

"To our amazement, we did it. We built the car! The learning experience of having actually taken an idea and made it real . . . incredible! It transformed everybody on that team. The car did all the things we said it was going to do.

"That introduced me to the real nature of engineering. It's an iterative process, trying to tweak it one click forward. It's always the pursuit of the ideal."

Dudley graduated in December of 1970 and a month later started with Ford as a product test engineer, feeling very sure that his star would quickly rise in that corporate world. After all, hadn't he done the impossible? He found out soon enough, however, that however dramatic his achievements in school, in the business world, things are done on corporate time. "There was not that sense of accomplishment that we had by coming out with that car. I used to spend months and months at Ford generating information on a project that I didn't understand. My job was to do testing, and I didn't know where it fit in the mix, whether I was making good information, bad information, whether it resulted in a conclusion or a solution. That was the fork in the road for me."

Dudley was fired in 1974 for insubordination. He'd been working seven days a week, ten hours a day, for over a year on material related to federal standards. The long hours on behalf of a project whose scope he didn't know were finally too much. "It was one of those classic scenes, in front of everybody. The big argument with the boss. He came down and it was 'Up yours!' and 'You can't talk to me that way' and 'I'll write a letter and put it in your file' and 'Screw you!' and 'You can't say that to me. You're fired!' and 'You can't fire me, I quit!'

"Eleven days later I was in Europe, backpacking around, sending real sarcastic postcards back to all my friends at the proving grounds, telling them to bail out. It was a bad attitude, bad mental attitude."

*   *   *

Dudley returned to the United States and after a while decided to pursue a course of study at Cal Poly in San Luis Obispo. To qualify for resident tuition, he had to live in California for a year before enrolling. He put all his worldly goods into a van, and he and his wife, Linda, moved to California with what he thought at the time was enough to let him sit out his year.

"We lived in Carmel. It was outrageously expensive. Now, you can live almost anywhere in the United States for a year for x amount of money. You can live in Carmel for about a week and a half on that."

Nearing the end of his bankroll, but with months to go to finish his year of residency, he was forced to look for work. He heard that a Santa Clara firm named Polymetrics was looking for someone to do research on reverse osmosis seawater desalting equipment.

Not having the vaguest idea in the world what that was all about, but needing the job and figuring there probably wasn't going to be much competition for that position, Dudley spent four days reading up on reverse osmosis, or RO.

"By the time of the interview I talked a pretty good RO story. I knew all the buzzwords and knew all about silt density indexes and transport phenomena across a membrane and all this jazz. I was really tap dancing in there. I was about the hottest RO guy that they had ever seen, however many of them there are flopping around out there. They said, 'This guy's got it!' "

While certainly not considered a subset of high technology, reverse osmosis consists of pushing seawater against a membrane that lets only the water, not the salt, pass through. Polymetrics built huge systems that were capable of purifying half a million gallons a day, to service large industrial and municipal installations.

Possessing the entrepreneur's insatiable quest for a business opportunity, Dudley soon saw at Polymetrics a poten-

Roy Dudley

tially huge market for smaller systems designed for yachts, remote resorts, small villages. When he approached the management there with his idea, they passed on it, saying they preferred to stay in the large-scale business they knew.

"But the idea wouldn't go away. It just seemed that somebody should do this. I think that's important. There are ideas that don't go away, ones you should listen to. The idea was to build a machine to make drinking water out of the sea real fast, real cheap. Two hundred gallons a day in a size that a man could literally bicycle-pedal to generate the power. It was a real compelling and romantic idea. And the fact that it was right at the height of the California drought didn't hurt. High on our own adrenaline, another guy and I put together a business plan. It had no cash-flow projections, no pro forma return on investment, none of the real technically correct ways to write a business plan. It was more of a romantic narrative of how we had this vision. It was intuitively obvious that it would be a great success; all we needed was five thousand dollars. Well, the business climate around San Francisco gives rise to people that tend to respond to that kind of thing. We got five grand on a poetic business plan!"

Dudley went through the $5,000 building a working prototype machine. When that amount gave out—and with no more coming in—he began to live off his bank cards. "We were hot!" When he was through, he had designed and built a small, portable box, using RO, that could take water out of the polluted San Francisco Bay and purify it within minutes to World Health Organization standards for drinking water. In addition, by fine-tuning the machine, he could make that Bay water pure enough for medical injections. All this during the California drought of the mid-1970s.

"We had one that worked, but we had no banking relationships, we were emotionally broke, financially broke. At night you'd hear your wife say things like 'Tell me again about how really neat this idea is.' It got pretty grim. We didn't know what to do. We just flat didn't know what to do. We were going to fold it all up."

Roy Dudley

By complete and utter coincidence, Dudley soon thereafter shared an airplane seat with the San Francisco bureau chief of *Business Week*, to whom he told his story. In February 1978, a full-page article on his system appeared in that publication. Within a few weeks, he received inquiries from over sixty countries. A delegation showed up from Saudi Arabia to escort him back to the kingdom for a demonstration. Talk about the power of the press.

"It was red-carpet time. It was just deluxe. We flew into Jedda and were caravaned in four cars, with the horns honking and all this stuff, out to the sheik's house on the Red Sea. I throw the hose in the Red Sea and seconds later this man tasted fresh water. It freaked him out! Water is part of the cultural fabric of Saudi Arabia. One of the original meanings of the word 'sheik' is 'he who owns the well.' They just went to the moon! So almost literally on the spot the guy says, 'Okay, I want the exclusive distribution rights for the Kingdom of Saudia Arabia. We're going to take this thing and make it big.' He gave me an initial order for five hundred machines, an open letter of credit for half a million dollars and an open purchase order that extended out for two and a half years. The whole value of this thing was over two and a half million dollars, and we said, 'All right, we'll start delivering in . . . pick a big number . . . a hundred and twenty days.' We thought that was a big number. Four months. I'd built an electric car with some guys in college in four months, certainly we can build a complete manufacturing and international field service corporation delivering high technology to Saudi Arabia within four months, right? And so we took the order, like fools."

Several businessmen in Australia heard about Dudley and his fabulous water machine and invited him to stop there on his way home from Saudi Arabia. "We had our one traveling watermaker. The goose that was literally laying the golden eggs. Our demo was a big media extravaganza at the Sydney Opera House. I was the young Yankee inventor. It was such a

Roy Dudley

circus, I can't believe it. Well, they doubled the Saudi order. We were signing all these exclusive distribution orders. I felt like the Pope in the thirteen hundreds, cutting up the world. 'This part is for you and this part is for you.' These letters of credit were coming in. We got these millions of dollars in orders. 'Let's go home and go to the Ferrari store.' ''

Upon returning to San Francisco, Dudley soon realized that even a once-in-a-lifetime opportunity like this could disappear as quickly as it came. He now had four months in which to create an international manufacturing and marketing operation, on letters of credit. Because he wanted to hold on to as much of the company as he could, and not dilute ownership by bringing in outside investors, he decided to job out the assembly work to what are called contract manufacturers. That was an abysmal failure, he soon found out. The manufacturer he contracted with knew nothing about Dudley's technology.

While tied up looking for someone to make his product in quantity, Dudley saw the first letters of credit begin to burn up. Orders kept coming in the front door, but evaporating out the back door because there was still nowhere to get the product made. "You can definitely do that, but you can't do it indefinitely."

He'd have to bite the bullet, he realized, and build his own factory. But factories cost money, and the place to get money in the Bay Area is from a venture capitalist, even if that means dilution of ownership. Better to own a smaller percentage of something than all of nothing, he figured.

He quickly raised $1 million and started Allied Water Corporation in South San Francisco, to make and market his SweetWater System.

His education as an engineer had begun to pay off. His education as a businessman was just beginning.

"I found out the guys that start companies typically are not the guys that end up running them. There's entrepreneurs and there's administrators. And it's the nature of entrepreneurs that they are risk takers. They are willing to roll the bones time and time again and play 'bet the company.' They also tend to be right most of the time. They have to be if they want to survive. Somebody has done an analysis, and in the decision-making process in that first year and a half you have to be right ninety-two percent of the time, because the decisions you are making are ones that are do-or-die. That willingness to bet it all, time and time again, is very attractive to an investor, a venture capitalist, at the low end. They want guys that have that kind of mentality, that move fast.''

But once things start to come together, once the company begins to acquire some value, the ''very qualities that were attractive at the outset become a threat. You do not want a management team sitting in a two-year-old company now worth two or three million bucks saying things every third day like 'I got an idea, you guys. Let's do something new. Let's paint the box green instead.' There's almost always a transition that happens going from start-up to fully functioning, regularized corporation, that transition between the entrepreneurial start-up team and the ongoing administrative team.

''What we found the venture capitalists will do is say to the entrepreneurs, 'Here's ten résumés of Harvard and Stanford M.B.A.s. Pick one who you think you can get along with for the longest period of time before he fires you.'

''The art is trying to make the transition a smooth one. Don't bring in the administrators too soon. You have to insulate them from real-world crisis-management situations. But don't let the risk-oriented entrepreneurs stick around too long either. You see, as an entrepreneur, you have to do everything. You negotiate funding. You solder circuit boards. You take the product publicity photos yourself. You do it all yourself, or with a partner. You know just enough about so many things it makes you dangerous. Then when you get into an organization that starts to become regularized, you tend to get in other people's faces. You see where you can help the controller, the field service guy, the secretary, the president, and you find yourself running around telling everybody how

Roy Dudley

to do their job. It tends to get on people's nerves. That's usually when you want to take a hike, anyway. You want to go back and get that rush again.''

Allied Water hired a president, took over sixteen thousand square feet of factory space in South San Francisco, and eventually had a staff of nineteen. The company was later sold to an existing firm in Wisconsin that already had an established distribution system in Allied's market. The SweetWater product is still being sold, and, in fact, the term has become generic for that type of water purifier. Dudley still owns stock in Allied, although because it is not traded publicly, he cannot sell his shares on the stock market. His holdings in the waterworks are not liquid. And Dudley did, indeed, move on to get the rush all over again.

When he was at Allied, it drove Dudley to distraction trying to talk with Saudi Arabia over an old telex machine. There just had to be a way, he thought, to hook a telex to a computer or word processor keyboard. Naturally, he found a way and designed a little black box that allows existing office equipment to talk over domestic and international telex.

With his previous experience under his belt, Dudley applied what he'd learned and his new Mountain View company grew smoothly. Dudley had come to realize, as had the young geniuses at Apple and their brilliant marketing guru, Regis McKenna, that what sells technology products in the home and office is not the technology but the approachability.

"People do not want things sitting on their desk that look like a cross between Darth Vader and a Nikon camera. We coined a term for it—'techno-dread.' We did not want to come up with the Ectotronics Telex Interface Hybrid Enhancement Module, or some nonsense like that. We wanted to have something that they wouldn't be scared of. We basically said: friendly, convenient communications. What's that? Chatting, like we're doing right now. So Chat became the company name and it became the product name. What do you pick for a logo? We used a multicolored parrot and called him Chat. It worked at every level.''

A new-product news release to the trade press, showing the product with a parrot on it, established Chat Communications as a force in a new category of the telecommunications industry. Dudley and his partner, John Murphy, raised the money, hired development engineers and set up channels of distribution. Soon after, a larger corporation named Plantronics took over running the company, now called Plantronics Chat. Then Dudley and Murphy phased out of the company. They still have a significant financial interest in the business, and Dudley says, ''I feel good about letting them drive it around.''

Having seen several of his companies progress from start-up to stability, Dudley has observed some things about human nature, his own included.

''When you start a company, you don't have offices and you don't have regular hours. You stay at home or you meet in places where the work needs to be done. You lose track of dressing a certain way. You lose track of the impropriety of having a beer at eleven in the morning. It's flat-out fun. You get a ton of work done and you're living a life-style that is the way I think everybody would live if they had the opportunity. Then all of a sudden the thing takes off. You get some money in. You get an office. A secretary comes in to do filing and typing and cooks the coffee. She shows up on Monday morning. She's real uptight about her new job. It's a work environment for her. Then the bosses show up and they come in wearing shorts and T-shirts, laughing and carrying on . . . it's hard on the employees.

''My partner and I worked very, very hard for a long, long time and we didn't really understand the necessity to maintain certain illusions. The management team in some people's minds needs to be aloof, distant, conservative, apparently always deep in thought, not able to be bothered with human stuff. The aloofness and distance is an innate paranoia on the part of presidents and managers not to let employees find out

Roy Dudley

that they really don't know what the hell is going on. I'm not afraid to let people know. If I get into a situation where I don't know what's happening, I'll turn to the group and say, 'I haven't the slightest idea of what's going on here.' ''

The painful lesson Dudley learned was that his kind of candor is not appreciated by a lot of people.

"One day John and I went to lunch and coming back we passed a little army surplus store in Mountain View. They were selling silk Japanese kimonos, with dragons on the back. Beautiful! Eleven bucks apiece. Fantastic! We both got one. Turned around and they had army-issue World War II pith helmets. Eighteen bucks. Let's get 'em. So we came back to the building and we walked in and we had pith helmets on and we were wearing these kimonos and were having a great time. We went in our office and we sat down and went back to work making phone calls all over the country, talking to distributors and dealers and fielding customer service questions and things like that. Over the phone, right? No one can see what we have on. But the fact that we were wearing pith helmets and kimonos created a shock wave in the company. The bottom of the whole thing for the people there was, 'Look, if these guys aren't taking it seriously, why should we?' And it caused a complete dysfunction there. What could have been a routine thing became a Big Magilla. It became An Issue."

Whimsy, Dudley discovered, is not always appreciated in an office, even if the office is one's own.

"I think I'll be more conscious of this in the future. There are people that work for a living and they don't want to know whether the funding is going to happen. They don't want to know whether this thing is at risk. They don't care whether the order comes in. They just want to go and type a letter and get their paycheck and go home. I used to tell everybody, 'Look, man, you've got to go out and start a company.' Because I thought that everybody wanted to have the same rush that I was getting. It turns out a lot of people don't.

Roy Dudley

When you start a company and it starts to regularize, people draw together and create a culture. And it should not be disturbed. Part of that culture is the shared belief that the management is competent.''

Having moved on yet again, Dudley found the call of the entrepreneurial swamp still too strong to stay away from. This time, two men from the East Coast that Dudley had met earlier had started a company in Connecticut called AMBI, secured venture capital and designed an integrated voice and data set—''like an IBM PC that's been stuffed inside of a business phone''—that allowed simultaneous voice/data transmission over standard telephone lines. Even before AT&T. In order to help market their product, called an AmbiSet, they approached Dudley and asked him to work with them—from California—in the start-up phase, based on his experiences having taken companies up before.

''One of the values that I bring to a business like AMBI is telling them up front when I'm going to leave. The last thing that a company wants to do is hire somebody who is good in a very narrow window of time and then becomes dysfunctional. Because you still have to keep him around because he's an employee. Part of my deal with these guys is, 'Look, I'm going to tell you where I'm going to get you. Then I'm going to get you all the way over there and I'm going to take my money . . . a big chunk of money . . . and I'm going to leave. But here is what you are going to have at the other end.'

''It's like the old Sam Spade movies where he does his thing and it's a job well done and he goes back to his office and throws his hat on the hook, puts his feet up and waits for the phone to ring. I guess success is being able to go out and get another one. Failure would mean not being given admission to the cartoon anymore. That would be terrible.''

For all his experience bringing new companies up, Dudley expressed genuine surprise when asked how many jobs he thinks he's created.

Roy Dudley

"Oh, wow! I never even thought of that!! Well, let's see. There's a way that . . . directly, probably over two hundred. Indirectly, some multiple of that . . . attorney time, printers, outside vendors, contract manufacturing, things like that. Between us, John and I have managed to stir about four million in investments. It all got spent. So there were people working out there. . . . I think that's pretty neat to think about, how many jobs we created. . . ."

If Dudley ever finally tires of being a "start-up mechanic," he feels he's had enough experience to enable him to become a venture capitalist, where the biggest skill is knowing *who* to back, more than *what* to back.

"I can see where having done a couple of these would be good credentials for being a venture capitalist. I would give the guy with experience the money before I would give it to the guy with the education. I can see a piece of 'blue sky' coming a mile away.

"People tend to think of venture capitalists as being benevolent grandparents, writing a check to help the kids out. Wrong! They are pragmatic businessmen who are flinty-eyed. Technically, they are a bunch of poker players. You're sitting on one side of the table and you're trying to make your million. And they are trying to make their twenty million. Part of their deal is to value your company at a price which is generally astronomically lower than you think it is. So you enter into negotiations. You have a real aggressive business plan and you say that you can get from here to there in twelve months for half a million bucks. They know you can't do it for half a million, that you're going to get nine months out and you're going to run out of money. But in the long run, you've got great product, so they will give you the half million and they let you run for it. You get nine months out and you run out of money. 'Okay,' they say, 'second round. You're in trouble, man. You need another half million? Come on, you were going to make it. This time the shares are no longer worth two-fifty. They are worth a dollar seventy-five. But we're going to give you three-quarters of a million. We're going to give you plenty to get you there.' But now they own seventy percent of the company. I've never been manipulated in that way personally . . . our companies never have been. I'm speaking not from experience, but just from observation of some cases. Venture capitalists tend to be pretty damn pro. And I think they should be respected as such. It's business. It's business. Strictly business. If you can't stand the heat, get out of the kitchen.''

For all the words about millions of dollars, written in respected business publications and spoken of in schools of business, it all, according to Dudley, comes down to sweat, pluck and chutzpah.

"You can sit over at the Decathlon Club with your quiche and your chablis and you can talk about your latest third round of funding at the twenty-million level and how you're going to do a public underwriting. But the guts of the thing is when you're standing at the contract manufacturer's and he's got your finished goods on the floor, but he won't release the product until you pay him the twenty-eight thousand that you have owed him for ninety days. You start thinking, 'If I can just get this guy in the other room while my partner comes in the back door and takes the product and delivers it to get the money to pay him off, it will all be okay later.' And you start doing whatever it takes to achieve the goal. I think that's closer to where you get your blood up. I think that's closer to real life. Of course, in the back of your mind you're hoping that whatever you do today will someday result in you, too, being able to sit in the Decathlon Club and talk about your latest forty-million-dollar deal. People do that all day long here, every day. That's what makes the Valley go round.''

6 | Robert Noyce

No individual's story charts the genesis of enterprising technical activity in Silicon Valley better than that of Robert Noyce. He has been at the center of this high-technology community and its spin-off and start-up activity from the beginning.

He was one of the earliest employees at the Shockley Semiconducor Laboratory, which opened in Palo Alto in 1955 and was the first semiconductor company to locate in the Santa Clara Valley. (There were already a few technology companies here then, dating back to the 1930s and '40s, such as Hewlett-Packard, Varian Associates and Ampex, but it was the launching of the semiconductor industry here in the mid-1950s that gave this place its nickname and marked the beginning of the explosive growth pattern.)

In 1957. Noyce was one of eight men who left Shockley to launch Fairchild Semiconductor, the still-existing granddaddy of them all, from which scores of companies, such as Frank

Robert Noyce

Deverse's IMI, would later spin off. While at Fairchild, Noyce invented the integrated circuit.

In 1968, along with Gordon Moore, Noyce left Fairchild to start Intel, the company from which would soon come the microprocessor.

His association with technical innovation and entrepreneurship hasn't diminished at all, as much as he might sometimes wish it would. He has seen a number of Intel employees go off to start their own companies, often in direct competition with Intel.

But before beginning Noyce's story of the development of Fairchild and Intel, some background about the Shockley Semiconductor Laboratory—and its founder, William Shockley, co-inventor of the world's first semiconductor—will be helpful.

William Shockley spent part of his early childhood in Palo Alto, while his father taught mining engineering at Stanford. He moved away from the Bay Area as a young man, and in 1936 received a doctoral degree from MIT. His thesis had to do with a then obscure field of study called solid-state physics.

Soon thereafter, having joined the technical staff of the Bell Laboratories of AT&T, Shockley found out that the Bell System was considering implementing telephone exchanges in which the circuits would be switched electronically, not mechanically as they were at the time.

In the late 1930s, however, electronic switching meant relying on glass-enclosed vacuum tubes. And although the tubes represented a great improvement over mechanical switching devices, they were a far from perfect solution. They required a warm-up time; they drew a lot of power; they burned out; and, because they were made of glass, they were very fragile.

Before the Bell System could even move on to tube technology, Shockley made a conceptual leap and began to think about, and look forward to, the possibility of switching

Robert Noyce

devices which depended upon solid-state, not vacuum-state, physics.

While metals such as copper and aluminum and silver are excellent conductors of electricity, and while wood and rubber are good insulators, there is a class of materials in between, such as silicon, germanium and gallium arsenide, which are semiconductive. That is, in their pure form they act as a sort of insulator; they tend to resist the free flow of electrons. However, adding impurities or "doping" these materials—for instance, adding small amounts of boron or phosphorus to silicon—makes them conductive. So doping a channel in pure silicon creates a high-conductivity pathway in a low-conductivity medium.

World War II put an end to Shockley's research into making a practical amplifying and switching device out of this observed physical phenomenon. But after the war he returned to the Bell Labs in Murray Hill, New Jersey, and in late 1947 he and his colleagues, John Bardeen and Walter Brattain, discovered the "transistor effect." They found they were able to create the amplification of an electronic signal passing through a semiconductor device. And therein lay the power and potential of electronics: the amplification of a micro signal to create macro effects, whether in a telephone switching device or the guidance system of a lunar probe. Shockley, Bardeen and Brattain shared the Nobel Prize for Physics in 1956 for this discovery.

This first working solid-state device was called a transistor. As is often the case with breakthrough technologies, its potential value was not fully appreciated at first, since the first transistors had very modest capabilities. The earliest commercial applications were in such mundane devices as hearing aids and portable radios that came to be nicknamed after the component.

The transistor (the electronic device, not the radio) overcame all the drawbacks of the vacuum tube: it required no warm-up; it used much less energy; there was less heat to

Robert Noyce

dissipate; it was smaller and more reliable and, being solid, wouldn't break. And so eventually before scores of applications opened up for the transistor, the most significant of which was in a new machine, growing out of the war effort, called the computer.

To many, even today, semiconductors and computers all seem cut from the same high-tech cloth; they're all of a single, alien piece. In fact, electrical engineering and computer science are two very different disciplines, and each had developed along its own course until, in the 1940s, the existing vacuum tube technology was used to power the huge numerical calculators needed for the war effort. The most famous of these machines was the now-legendary ENIAC— the "electronic numerical integrator and calculator"—which was designed by J. Presper Eckert, Jr., and John Mauchly at the Moore School of Electrical Engineering at the University of Pennsylvania.

With the end of the war, the productive interdependence between electronics and computer science increased exponentially. The Cold War arms race, the postwar information explosion in the sciences, the emergence of a peacetime consumer society, and the growth of government involvement in citizen welfare all made the need for fast, accurate "number crunching" and information flow even more apparent and pressing.

And so it was that the acceptance of transistors to replace vacuum tubes in the early 1950s was well timed, indeed, if the computer was ever going to advance beyond the temperamental, cumbersome behemoth it started out to be. (The night ENIAC was first switched on, it drew so much power that the lights in the surrounding community dimmed.)

As more transistors became available during the '50s, applications for them grew—and not just in computers and telephone switching systems and portable radios, but in tape recorders, intercoms, radar, even the guidance systems for chicken-feeding carts.

Robert Noyce

New applications demanded more product. In 1954, 1.3 million transistors were produced; by 1957, the output was up to 30 million devices a year. The economy of scale in such mass production cut the price dramatically. In 1953, a single transistor cost $21; by 1958, they were down to $1.50 each.[3]

Realizing the business potential in what he had helped develop, Shockley left the Bell Labs in 1954 to start his own company to make solid-state devices. Dr. Arnold Beckman, the educator/engineer/businessman and founder of Beckman Instruments in Fullerton, California, agreed to back Shockley, who chose to open his new company in Palo Alto: for its proximity to Stanford, with its well-trained engineering graduates, for the congenial nature of the place itself, and perhaps because it was the setting of happy childhood memories.

That Shockley was an innovative genius has never been in question, whether his work is considered as pure, applied, basic, practical, industrial or fundamental research. That Shockley could attract and bring together a pool of extremely talented, energetic men in Palo Alto is also beyond dispute. But that Shockley could manage the forces he'd brought together in his new company is another matter. Judged against some of the later successes in Silicon Valley, the story of the Shockley Semiconductor Laboratory, as a business entity, is not that impressive. The company was sold to Clevite Corp. in 1960, which in turn was sold to ITT in 1965; then ITT left the semiconductor business in 1968. On the other hand, if William Shockley hadn't set up shop in Palo Alto, and brought together there people like Robert Noyce, there might never have been a Silicon Valley on the San Francisco Peninsula at all.

Very few engineers—in Silicon Valley or anywhere—have been as resoundingly successful as Robert Noyce. However much he may claim to have backed into one role, he has secured a preeminent position, both as inventor (of the integrated circuit) and as businessman (cofounder of Intel). Yet if there is any intimidation in meeting him, it is in the anticipation only. A short man with a brisk stride, he moves through a warren of office cubicles with a pace that demands the visitor run to keep up. But once engaged in a conversation in his own, unprivate cubicle, undifferentiated from the hundred or more around it, he is as informal as his tieless, open shirt collar appearance indicates.

The informality, genuine and not studied, still masks a very methodical individual. On a notepad in front of him he outlines every question he is asked, to pursue ideas in the order in which they were raised.

After studying at Grinnell College in his native Iowa, Noyce, the son of a minister, received his advanced degree in physics in 1953 from MIT. He recalls being introduced to William Shockley at a conference on solid-state devices at which he, Noyce, was reading a paper. Soon thereafter, he was invited by Shockley to join the new organization in Palo Alto.

"Shockley was trying very hard to be innovative in his management style. Some of those things backfired, some of them worked well. But overall I would say that he alienated a number of his staff. They, and I was not part of that group, went to Arnold Beckman and asked him if he could do something about it. I guess I would have to say that what Arnold did was too little too late. And so the situation blew up. We had an in-house rebellion."

Robert Noyce and seven others—Jean Hoerni, Eugene Kleiner, Gordon Moore, Jay Last, Victor Grinich, Sheldon Roberts and Julius Blank—decided to leave Shockley and go out on their own. If at all possible, they wanted to stay together as a group to start another semiconductor company, but they needed financial backing to do that.

The father of one of these young men had an account with the New York investment firm of Hayden, Stone & Co. and mentioned his son's plight to someone there. Soon after that,

Robert Noyce

a partner at Hayden, Stone and a young executive there named Arthur Rock flew to California, met with the eight, and decided to work with them to secure financing for their proposed new venture.

Rock went to a number of companies to try to interest them in backing the project, but was turned down at every stop. The concept of capitalizing new technology ventures was virtually unheard-of then, and in fact the term "venture capital" itself wouldn't be coined until 1965. By Arthur Rock.

Finally, Rock approached Fairchild Camera and Instrument in Syosset, New York. That company's founder, a creative, multitalented man named Sherman Mills Fairchild, was interested in the project, and his company loaned the new group $1.5 million.

Fairchild would finance the new company, but it was to be owned by the eight founders, plus Hayden, Stone and, later, Dr. Ewart Baldwin, who was brought in as general manager, since none of the founders had business experience.

In return for putting up the capital, however, Fairchild was granted an option to buy the company. That was executed in 1959 for a purchase price of $3 million.

And with that purchase, the eight former Shockley employees-turned-entrepreneurs became, in effect, employees again, of what would be called the Semiconductor Division of Fairchild Camera and Instrument. Alas for Fairchild, that wasn't a situation that would last for long. Once bitten by entrepreneurship, many found it hard to be employees again. Ironically, the last in was the first out. Baldwin, the man brought in as business manager, left before the buy-out to form Rheem Semiconductor. History doesn't record his last words at Fairchild, but they might well have been "After me, the deluge." In any case, his departure left an administrative vacuum at Fairchild, and Robert Noyce grudgingly began his own transformation from engineer to business manager.

"They asked me to take that administrative job and I said,

Robert Noyce

'Well, I'll do it until we find somebody.' I was not particularly interested in it. I don't think I had the self-confidence. But we looked for a couple of months and then I finally went back and I said that we hadn't seen anybody that could do the job better, so I guess I'll go ahead and do it. I remember talking to a Harvard Business School graduate, saying that really I don't know much about this business. He said, 'Come on, you'll learn.' '' A pause, then Noyce adds, ''I did . . . some, anyway.'' And he flashes a disarming grin, worthy of any celebrity.

And as quickly it disappears, and he again assumes a neutral expression I am more used to seeing in rural coffee shops in the Midwest: absolute alertness with no hint of what's really being mulled over. For all his informality, one has the sense with Noyce of being in the presence of a man of notable contributions to science, as well as business, a role he grew into gradually and gracefully.

''I found that it was interesting being at the focus of the information network, if you will. At least in a smaller organization like that, people are coming to you for decisions of one sort or another and I had a much better picture of what was going on from that position than from the previous position, directing research activities there.''

While at Fairchild, Noyce invented the integrated circuit, a type of semiconductor that incorporated several transistors—ten or twelve then; up to half a million now—on a single piece of silicon.

''Co-inventor would be a more accurate title. Jack Kilby at Texas Instruments and I have been sort of jointly credited with that. It was totally independent work. He was pursuing one line of thought. I was pursuing another.''

As the technology was imploding, packing more power into less space, the talent at Fairchild was exploding, following Baldwin's example. Hoerni, Last, Roberts and Kleiner

Robert Noyce

left Fairchild in 1961 to form Amelco. Six others from the company then went off to form Signetics that year. In 1962, other "Fairchildren" formed Molectro. In 1963, General Micro-Electronics. In 1967, National Semiconductor. That same year, Hoerni started yet another company, Intersil. By 1980, nearly seventy firms would be able to trace their lineage directly back to Fairchild Semiconductor.

The parent just could not seem to hang on to its children. Yet Noyce stayed on. Having watched most of his colleagues depart between 1959 and 1968, Valley watchers wondered when, or if, Noyce would ever join the throng. Or, having done it once, would he try it again?

"Fairchild was going through a great deal of management shakeup in New York, and it just didn't seem to be working. It was an East Coast management style and a West Coast management style, and they're quite different. One is much more autocratic than the other. This environment here is really quite thoroughly democratized in its management style.

"The other motive for my leaving was that I did see some opportunities in being able to get on with the new technology. So it was backing away from the big administrative task, and back again into the technology and worrying about project direction. Then, also, it was to prove that the first one wasn't just a lucky break . . . to see if I could do it again. And frankly, another motivation was that everybody that had walked away from Fairchild was richer than I was."

In 1968, Noyce, Gordon Moore and another Fairchild alumnus, Andy Grove, formed Intel. The venture backing, once again, was secured through Arthur Rock. The company was formed with the intention of turning away from the dominant memory-storage technology of the time, called "core," and instead producing the first semiconductors used for memory. The integrated circuit was now used not only for logic operations but as a memory device as well. Soon after, engineers at Intel developed what is, possibly, the most significant invention of our age.

"A small company in Japan came to us very early in the history of our company and said, 'We want you to build this set of custom chips for our calculator.' At the time, we had a grand total of, I think, about thirteen engineers in the whole company. To do the job they wanted would have required a high percentage of our staff, and we didn't want to invest in that. We had other things to do. Then Ted Hoff came up with an idea. We'd build a small computer on a collection of four chips and program them to look like a calculator. And that idea was presented to the Japanese and they accepted it. We had seen it as a simple way to do their task. We could do their entire family with one set of chips, like square-root functions or whatever. Then we realized that not only could it do a bunch of different calculations, but it could do any of the standard control functions, as well," like drive the keyboard or a peripheral storage device.

The capabilities of that microprogrammable central processing unit on a set of chips—called the MCS-4—was later integrated into a single, advanced form of integrated circuit—the 8008—and soon came to be known by a simpler name: the microprocessor. In addition to Ted Hoff, the names of Stan Mazor and Federico Faggin are also on the patent for the first such device. That superchip represented the convergence of electronics and information processing: the component *is* the computer; the computer is a single component. The microprocessor would soon make possible the microcomputer, which is as big as it is only to accommodate us. We'd have a hard time getting information into or out of a microprocessor without a keyboard, a printer and a terminal.

Noyce's involvement with entrepreneurs hasn't lessened since forming Intel. In some cases, the entrepreneurial activity has helped Intel; in some situations, it hasn't. In the former category would be Silicon Valley's most famous start-up.

Robert Noyce

"Mike Markkula had been marketing manager here. He made quite a bit of money out of Intel by going into hock and using every penny he had to buy Intel stock when there was real risk in doing that.

"He resigned in the mid-'70s and was looking around for investment opportunities. That's when he ran into the two Steves [Jobs and Wozniak] and joined up with them and was the first significant outside capital that came into Apple. We were the genesis of Apple in another area, of course, and that was originating the microprocessor, which was key to the microcomputer's development."

Other start-ups, however, have not been so affectionately remembered by Intel, particularly those initiated by employees who went into competition with their former employer. "I think there is an enormous difference between starting a really new venture in a new field, like Apple, and starting a 'me too' venture. And some of the 'me too' ventures have weakened the existing companies. I think it is of questionable ethics to conspire while you are employed by someone to do something else. It's much cleaner to leave and then go to work on it. And we've had a couple of cases at Intel of the former which we have been upset about. For example, we've had people working on business plans with our market data, and recruiting workers while still employed by us, recruiting within the company. I feel those are unethical practices."

They no doubt are, but didn't Noyce himself go off and start new enterprises, twice—leaving Shockley to start Fairchild, then leaving there to form Intel? Is entrepreneurship only valid while one is an entrepreneur, then predatory when one has territory to protect?

"You even ask yourself that question when this sort of thing happens. The only thing I can say in our defense is that in the Shockley case, it was after we consulted with management. And Fairchild did nothing on diodes [which were being done at Shockley] for several years thereafter. In the case of Intel, I resigned before I did anything. And Fairchild was not working on memories at that time, so it was a different field. In both cases, the new activity was something different than what had been done in the old activity. It's when the new activity is the same that it gets to be a question of ethics.''

Having seen so many companies start up, what general characteristics, what personality traits, has Noyce observed about the breed?

"I think most entrepreneurs don't see the risk as very high. They are people with high levels of self-confidence and they are people whose skills are in high demand, so that even if the venture should fail they wouldn't starve. They may take two years of advancement out of their current career, and have to start back where they were then, but I don't think that even that really happens.

"Very few people, at least when they are starting out, have the specter of failure in front of them. They're just blind to that part of it. Maybe it's self-preservation. You're looking at those risks and you're trying to figure out how to avoid them, but you don't really think you'll ever fall off the cliff. I mean, why does one climb a mountain? Because it's there. It could also be because there's a risk of falling off the damn thing, and that's part of what gets the adrenaline going. I don't think that the risk of failure is a deterrent. Obviously, intelligent people are not going to go out and try to do something which they are convinced will not work.

"The problems of success are handled by different people in different ways. It has cost a lot of people their families in the Valley. I'm divorced, remarried. I'm not sure that being deeply involved in a venture would have affected that one way or the other, but it is a fact.''

How does Noyce deal with his own success? Wealth and fame aren't usually what one expects when beginning a career as a physicist. "Financial success is sort of astounding to me. I never expected to be rich. And then you suddenly look back and say, 'I could stop working for the rest of my life.' I guess

Robert Noyce

that is the point at which you say, 'Well, okay, I've reached financial success, security. But I don't stop working because that's the fun in life.'

"The greatest impact in terms of fame came with my election to the National Academy of Engineering and the National Academy of Science, and being given the Medal of Honor by the Institute of Electrical and Electronic Engineers. Some of those things sort of came in rapid succession in the mid to late seventies. 'Why are you giving me this?' I wondered. Do I enjoy it? Am I embarrassed by it? Oh, a little bit of both. I suppose I play to the crowds in that. Everyone likes to be recognized, I believe."

However much Noyce has come to terms with his reputation, he is remarkably self-effacing in discussing his technical accomplishments. "The fact that the integrated circuit was done at two places, essentially simultaneously, says that if neither Kilby nor I existed, somebody else would have done it within a year or two. So it may be just pushing the pace a bit, but these were ideas whose time had come."

And while the stereotypic engineer is sometimes thought to have no concern over the effect of his technology on society, Noyce is very much concerned about that interaction.

"I've been worried about the two cultures [technical and liberal arts] getting further apart. And it's because of the fear of the new technology displacing people who are not familiar with it. But I think that will cure itself as the young of today get into society. Because they're going to be familiar with all this.

"The role of technology in eliminating jobs is a concern that hit Europe much earlier than it hit America. And I think they answered it in the sense that they realized more jobs would be lost if the new technology is not adopted than if it is. There clearly will never be as many steelworkers or auto workers as there used to be. But there are always needs in the society to be filled, and it will take foresight and insight to

Robert Noyce

recognize what those are and transfer that labor into those new needs. We're getting short on housing now. I would think that there's a significant amount of labor that could be absorbed there. We've seen the highway programs become a need of the society. We have a lot more to do in education. The difficulty will be to fit the available excess labor into the new needs. And that is my idea of a very difficult task.''

7 | Alan Shugart

Entrepreneurs have a reputation for being very verbal people. They have to be, to take a private vision and make it a public reality. Engineers, on the other hand, are often perceived as being nonverbal. Something works or not, and all the words in the world won't make it fly if it's not designed right.

The people profiled so far, though each has a background as an engineer, responded as entrepreneurs to questions put to them; that is to say, they responded at some length. However, one of Silicon Valley's best-known, most highly regarded engineer/entrepreneurs, Alan Shugart, said he would prefer to be asked only questions he could answer with a yes or no. Once the conversation began, however, he spoke at somewhat greater length, but he really does prefer to keep a low profile. "I'm a low-key individual. I don't like publicity. That kind of thought process escapes me. But I think it's good for the company for me to be accessible to writers."

Dressed in a sport shirt, in his spacious office looking out

Alan Shugart

on a redwood forest, Alan Shugart looks more like a working engineer than a corporate honcho. He grew up during the Depression, and graduated from the University of Redlands in Southern California in 1951 with a B.S. degree in engineering/physics, one of only two such majors in his class at that liberal-arts school.

He went to work for IBM in Santa Monica as a customer engineer, fixing broken accounting machines, sorters, and key punches in the days before computers. IBM eventually transferred him to a research and development lab in San Jose.

In the middle to late 1950s, that lab developed the concept of a random access memory, allowing data to be stored on a rotating disk surface and retrieved without requiring a scan of all nonrelevant information first. It was like being able to set the needle down on a specific track of a record to listen to just that cut, as opposed to playing ahead in fast forward and then jockeying around to find the desired spot on a tape.

The 24-inch random access disk drive mechanism that came from IBM in the late 1950s greatly augmented and simplified the computer's ability to store and access massive amounts of information.

Shugart worked on further memory technologies at IBM throughout the 1960s, including some of the early development work on a type of random access memory device now called a Winchester disk drive—a stack of rigid platters in a sealed environment, storing large amounts of information that are readily accessible by a host computer. He might have been content to remain at IBM, which he still considers "the best-managed big company in the world," except that the corporation forced his hand.

"I was transferred to New York, but I never moved. In my job I had responsibilities in a lot of different labs and would visit them. I would always just make sure that my Monday visit was in San Jose. Then I would fly from New York to San Jose on Friday night, spend the weekend here, then fly

Alan Shugart

from San Francisco back to New York on the Monday-night Red-Eye. I didn't want to move to New York.

"I decided to look on the outside. I hadn't ever worked anyplace else. Hadn't even looked. I discovered there were a lot of opportunities that never occurred to me. So I quit IBM in 1969, without a job."

He joined Memorex as vice-president of product development, and remained there for over three years. Then, with several others, he started Shugart Associates in 1973, to make "floppy" disk drives, now also called "Shugart drives."

"I had the big name and all, so I was the boss. But the equity split between the nine of us was nearly equal. I worked there for two years and left in a huff after a big argument with the venture capitalists over who was going to put in more money and at what kind of equity participation. And as a result of a disagreement, I quit. I'm not sure . . . if I hadn't quit I would have gotten fired. So I don't know if I quit or got fired."

It can't be easy to leave a company that carries your name, but apparently that didn't bother Shugart. He planned to get out of the business anyway.

"I started salmon fishing. I had a commercial license and my own boat. Never made any money at it, but I did it anyway. Paid for beer."

And how, one wonders, does he compare being president of a high-technology company with salmon fishing?

"You get a suntan when you're fishing."

There was no "golden handshake"—no big payoff—for Shugart when he quit Shugart Associates. No payment even for the right to continue using his name. "I was flat broke. I collected unemployment for several months."

Eventually, the novelty of salmon fishing wore off and Shugart took up independent consulting and "general horsing around."

Alan Shugart

In 1979, Finis Conner, a cofounder of Shugart Associates, quit that company. He served for a while as marketing vice-president at another company, then quit that position and approached his former colleague with an idea. With all the desktop microcomputers beginning to appear in the world, many of them with several floppy disk drives attached, there would surely be a market for a micro version of the higher-capacity Winchester drive which Shugart had worked on years before. Consider a product offering fifteen times the memory capacity of a floppy-type drive, occupying the same amount of space and costing only three times as much. Shugart agreed with Conner, that the product would be a natural. And being the first ones with the idea, they had an open marketplace.

"The name of the company, when we started, was Shugart Technology, because I had previously incorporated the company and the name about a year earlier. It had never become active, but the corporation existed, so we went ahead and used it knowing that at some point we might change our name because we didn't want to be confused with Shugart Associates, and we knew they wouldn't like it anyway. I was looking for a name that would replace Shugart, with seven letters and starting with an S and ending with a T and with a G in the middle. Seagate was the closest thing I came to."

Seagate Technology is located in the beautiful, forested Scotts Valley, on the seaward side of the Santa Cruz Mountains, away from Silicon Valley. That part of Santa Cruz County, with its wooded mountainsides and access to Monterey Bay, has always been a desirable place in which to live, but the area has traditionally suffered high unemployment. Shugart and Conner decided to locate their new enterprise there because they both lived on that side of the mountains.

Seagate's headquarters complex, seen through an opening in the forest by a motorist passing on Highway 17, could be mistaken for a luxury resort or prestige condominium develop-

Alan Shugart

ment. Leaving the freeway and coming into the complex on Disc Drive, one sees that the whole place is still in the process of coming to be, with building going on everywhere.

Seagate's growth boggles the mind, even by Silicon Valley standards, with sales of $9 million in 1981, $40 million in 1982, $110 million in 1983, and well over double that in 1984.

And as for competing with a company that still bears his own name?

"It used to bother me in the early days when I would hear people complaining about Shugart Associates. People would say they didn't like something. I'd hear it used with bad adjectives. It bothered me a little because it was my name. Shugart. Now I see it in the paper all the time and it doesn't bother me. Don't even recognize it as my name, I guess."

Whatever discomfort Shugart felt seeing his name taken in vain, it's made up for, he feels, when he reflects on his role in his industry and in society. He sees himself as "CEO of a four-year-old company that is doing well, providing jobs. We pay for a lot of people's educations. We don't have smoke-stacks polluting the air or dumping chemicals into the ground. We've advanced the power of desktop computing by two years so companies can be more productive, hire more people, be greater assets to society. I'm pretty proud of that. And seeing my friends get rich. We've made twenty or thirty millionaires here. I help my friends. I help companies get started. I cosign loans. Act as a venture capitalist by default." Yet while he feels a sense of self-satisfaction, the visibility of his role is an embarrassment.

"I'm a guy that likes to sit back in a corner and watch people. I don't like to be the center of attention. It bothers me a lot at trade shows like the National Computer Conference and Comdex, so I don't spend a lot of time at them. I went to the Consumer Electronics Show last January in Las Vegas and it was the neatest show I have ever been to. Nobody knew me! I just walked down the aisles and I could go up to booths and talk to people about their products and nobody bothered me.''

That's one of the reasons why, unlike most of his professional peers, he enjoys getting away from the computer industry in his spare time. For him, ''getting away'' is not going to a professional gathering in another city. Instead it's going to a neighborhood bar, because ''there's not a single soul in there that even knows what a computer is. And that's really neat. I think if you're not playing in a different environment, you're not playing.''

It's uncommon to find someone in his position able to keep one foot down home and the other in high tech. Shugart seems to fit in both worlds with equal facility. When asked what professional organizations he belongs to, he replies, deadpan, ''A country club in Carmel.''

Shugart sees himself as a ''survivor,'' and one so confident about surviving that he doesn't remember the hard times. Except one: ''Collecting unemployment. You don't do that because you like to appear in line.''

When asked if he thinks the world breaks down into two cultures, technical and humanist, he says, ''No . . . three: clerks, poets and soldiers. But it's a continuous flux. There used to be a farming culture. That's a very small part now. There was an industrialist/factory worker culture. And I think that's on the way down. Now it's the clerks, meaning the information-processing society. I don't know what the next change is. Poets and soldiers may become a subset of the whole information-processing society.''

Considering that this man's products, the floppy disk drive and the micro Winchester, have become the emblems of a new age, he has—perhaps because he's a child of the Depression era—a considerable respect for the basics.

''I think we need to go back to fundamentals. Our society got to the point of being fat, dumb and happy twenty years

Alan Shugart

ago. Kids learned how to enjoy life, as opposed to how to work. We've got to go back to fundamentals, so kids go back to school longer and learn better. Learn more work-related types of things. Less poetry and more clerks.''

Leaving the main building of the Seagate complex, the visitor sees before him a vista of forested mountainside that would assure the success of any resort lucky enough to be located here.

The sense of satisfaction Shugart says he feels, the physical and social environment which he and Finis Conner created here, is on the far side of the world from the entrepreneurial swamp. But I suspect that Roy Dudley and Alan Shugart, if they met, would find they have a great deal in common, though I doubt if there is a single word that describes what that common element is. And even if the word did exist, those who are unwilling to risk bankruptcy, burnout and going belly-up would probably never understand it.

3 | THE DEAL BACKERS

8 | Venture Capitalism

There has been a lot of discussion lately about the economy's metamorphosis from a sunset to a sunrise industrial base. Silicon Valley is often cited as a fulcrum point in that shift, and nothing in Silicon Valley better represents the phenomenon—literally—than does the view in the late afternoon from Interstate 680 in Fremont, the East Bay community across the water from Santa Clara.

Looking to the west from the freeway, the passing motorist sees nearby the mammoth, dark, brooding silhouette of what was once a major General Motors West Coast assembly facility that employed over 6,500 people. Back-lit by the setting sun, the now-unused plant seems like a dinosaur mired in a tarpit. And all around in the twilight sparkle the lights of dozens and scores of new enterprises, like Huey Lee's ATS, sprouting like so many mushrooms on the wide-open hillsides of the East Bay. In the offices and shop floors of these companies, the lights stay on, the managerial activity continues until eight, nine, eleven in the evening. And this isn't the night crew. These engineers and entrepreneurs will be back again at eight tomorrow morning to begin another fourteen-hour day.

Horses and sheep graze, crickets chirp on the nearly tree-less hillsides near the intersection of I-680 and the Mission Boulevard exit. And inside the tilt-up structures of the office parks, blooming as profusely as mustard flowers amid some remaining abandoned farm sheds, a revolution is taking place. One consequence of that revolution: the auto assembly plant will reopen, under joint GM-Toyota management, with highly robotized assembly operations.

This revolution rests on a very complex relationship that exists between two individuals: the electron-manipulating/deal-making entrepreneur and the deal-backing venture capitalist.

This partnership is not unique to Silicon Valley, but no-where is it more visible or does it have greater impact. This talent-money tango, based on hunch, persuasion, gut instinct, rapport or whatever, has led to the development here of such things as the integrated circuit and the microprocessor, de-vices which have opened new industries, and whose potential applications are still being developed.

While it took only $1.5 million to start Fairchild Semicon-ductor in 1957, because of the increased precision and smaller tolerances required in circuit manufacturing today it may require about $25 million in start-up financing to open a semiconductor foundry now. The enterprising engineer/business-man has to have the money in hand to bring together the resources he needs to make his first product. It is very difficult, though not impossible, to "bootstrap" a technology company, growing the business on its own earnings, when you need hundreds of thousands or millions of dollars' worth of hardware to assemble and test the widget in the first place.

For their part, the investors are drawn to these new compa-nies by the promise of unprecedented return on their investment. In early 1975, for example, the venture capital firm of Kleiner, Perkins, then of Menlo Park, invested $50,000 for 200,000

Venture Capitalism

shares of Tandem Computer, a manufacturer of computers for on-line transaction processing. In December 1977, Tandem's stock began to trade in the public market for $11.50 a share. Every dollar Kleiner, Perkins invested in the initial stage was worth $46 less than three years later. Clearly, there's a bond of mutual self-interest between the entrepreneur (in this case, James Treybig) and the "v.c."[4]

Venture capitalists and entrepreneurs differ in discussing their work. The latter are vocal, downright voluble, in talking about their company, its products and markets. In fact, there's nothing they'd rather talk about. Talk is the first step in making it real. Venture capitalists, on the other hand, are generally very quiet about their work. Theirs is a tightly knit fraternity in which flamboyance is a deadly sin, and the details of any single deal are treated with the same circumspection the priest reserves for confessional secrets.

At its best, the entrepreneur–venture capitalist relationship, when established, is amiable, professional and built on mutual respect. There are many people in both groups who speak with admiration bordering on awe for their partner.

But there are plenty of strong, negative feelings in both camps, too, about those on the opposite side. Some entrepreneurs are adamant in their refusal to seek venture financing, even if it means stymieing their company's growth, because they refuse to relinquish ownership of up to 70 percent of the business to their backers. Is it surprising that an individual who goes off on his own is individualist enough to want to retain as much control as he can of the enterprise he will sweat blood over?

On the other hand, there are entrepreneurs who know that not getting the appropriate investment in the company early on will choke it before it even gets started. They prefer to have a smaller slice of a larger pie, or, as one San Jose businessman put it: "The choice is between staying on the *Titanic* and remaining warm and cozy as long as possible, or getting into one of those cold, wet, noisy lifeboats.''

Some have tried to secure venture financing, to no avail—or have balked at the amount of the company they would have to relinquish for the backing—and in their frustration have coined the term "vulture capitalists." That same term is also coming to be used by the heads of mature, well-established technology companies, beneficiaries themselves of similar financing years before, who are livid that some venture practitioners today try to lure their best staff people away to set up competing organizations.

Even the entrepreneur who has secured financing occasionally lives in terror of his venture partner. In one case, the backers told the president—someone with a management, not a technical, background—that an engineering problem had to be solved by a given date or he was fired. Imagine the president's frustration: how can he "will" a technical breakthrough, especially when he's incapable of contributing to the work himself? But the man knew that however unreasonable the backers' request, it was theirs to make and his to accede to. As it turned out, the engineering team succeeded, but the president paid an awful price in alienating many of its members.

For their part, venture capitalists say they bring much more than money to the arrangement in order to help the first-time business leader survive the vicissitudes of starting up. They are—the better ones—experienced businessmen themselves who can advise the novice president about any number of things to avoid and aim for. They can suggest good support services to hire: accountants, lawyers, ad agencies. In fact, to have secured the backing of certain venture capitalists makes success almost inevitable. These certain backers have such excellent reputations for picking winners—in addition to being plugged into the most effective networks of contacts—that simply to have them behind the new company often gives it the legitimacy and clout of a Fortune 500 firm from the start.

Venture Capitalism

That means a lot if the new company is one of thirty entrants into a new industry and is in need of an instant reputation for solvency and respectability. Also, the fact that a certain v.c. is known to have "deep pockets"—is able to back a venture for a long time with patient money—intimidates those whose next round of financing is an open question. Industrial customers who buy in large quantities and demand follow-up service are prone to go with the company that seems most likely to stay around in the long term. And by giving their contracts to the company they perceive as becoming successful, they assure that company's success.

While the venture capitalist is often the one who is seen as being in the power position, it is reassuring to enterpreneurs to know that the financiers have problems just like everyone else. Though sometimes with a twist.

One Palo Alto venture capitalist was faced with the sudden departure of the president of one of the companies in his portfolio. Until a new president could be hired, the v.c. had to step in and run the company on a day-to-day basis, while continuing to run his own investing affairs. For most business people, coming home with the news that one is now the president of a company would elicit spousely congratulations. Not so in this case. The venture capitalist's wife, commiserated with her husband's added responsibilities by saying, "I'm so sorry, dear. I'm sure you won't be president for long."

If entrepreneurs are free to bad-mouth venture capitalists, the prerogative, of course, goes the other way as well. And while the number of entrepreneurs seems limitless, the number of venture backers is finite. There are perhaps ninety in Silicon Valley, and maybe five hundred nationwide. Their communications network is reputed to be instantaneous. And lethal if need be. The entrepreneur who is discovered trying to put something over on his backer or prospective backer—withholding some element of risk, or overstating the product's capabilities, for example—will not only find these negotiations terminated, but discover that the resulting bad rep will likely extinguish the chance of future deals anywhere else as well.

A given company represents one of several investments in the financier's portfolio. It means a lot to him, certainly, and many venture capitalists care personally about their entrepreneur associates. But ultimately, the financier is usually not involved on a day-to-day basis.

For the founder, however, the same company represents life itself—his or her life, or several years' worth of it, anyway: the time not available to spend with sons and daughters; the dinners he or she was too anxious or irascible to eat; the night sweats; the breakthroughs; the relationships severed irrevocably; and, ultimately, the unmitigated pride in having taken an idea and breathed life into it.

The relationship of the founder to his or her business/social entity is as complex and as deeply felt as the love-hate relationship Michelangelo had for the Sistine Chapel ceiling. And just as surely, there are venture capitalists who, like Pope Julius, stand back in awe and say, "I made it possible, but you did it."

Why has the field of venture capital suddenly become such a hot one, replacing consulting as a fashionable career path for the brightest graduates of the nation's leading business schools? People have been starting businesses for thousands of years. And, in fact, venture capitalism has been around for a long time, under other names. Isabella put up her jewels to back Columbus's venture across the Atlantic, and thereby became his patron.

Merchants in San Francisco and Sacramento grubstaked forty-niners, keeping a steady income from their stores while leaving open the possibility of a tremendous windfall if their miner hit paydirt. Betting on the come is as old as ambition.

In the past, however, investments in highly speculative

Venture Capitalism

enterprises were generally made by wealthy individuals or families, such as the Rockefellers (in Eastern Air Lines) or the Mellons (in Gulf Oil and Alcoa). And then, the investment represented a sideline to the main source of wealth.

It's only since World War II that the function of investing in speculative enterprises has become institutionalized. Companies have been set up and run by professionals on a full-time basis, with the sole purpose of raising pools of money to back new ventures. It has become an industry with its own trade associations, like the National Venture Capital Association, and a trade periodical, *Venture Capital Journal*. In fact, there are venture capital organizations now that help start new venture capital organizations. Such investing is no longer a private arrangement between individuals, although the deals are still largely based on personal rapport.

While not restricted to backing high-technology companies, the venture function has been most closely associated with those types of firms from the beginning. The first of these institutionalized firms, General Georges Doriot's American Research and Development Corp., which started in Boston in 1946, would later invest in, among others, today's number-two computer maker, Digital Equipment Corp.

Yet because of this link between venture capital and high technology, the phenomenon traces some of its deepest roots, and has enjoyed many of its most spectacular successes, in Santa Clara County. There is a compelling logic for the new technology companies to be so closely associated with the new form of financing. Where else is the technology entrepreneur to go? Bootstrapping the operation is usually too slow in this fast-moving world. Banks, as the conventional wisdom holds, make loans only to those who can prove they don't need them. How does someone with a new product, based on a new technology, aimed at a new market, satisfy the risk aversion of a banker?

Insurance companies and pension funds, with their collective assets of hundreds of billions of dollars, were, until recently, closed as sources on the grounds that a "prudent man" would not invest in high-risk high tech.

It took the leaders of early existing innovation companies, like Arnold Beckman of Beckman Instruments and Sherman Fairchild of Fairchild Camera and Instrument, to see the potential in these companies. And successful experience in arranging such support allowed a man like Arthur Rock to go from facilitating deals, as discussed in the profile of Robert Noyce, to the point where he is today his own source of venture funds.

Venture money (some 80 percent of it) is attracted to the high-technology industries for a number of reasons, but primarily because payoffs of ten to one in less than five years are not uncommon. No other industries are so consistently showing such returns, in spite of cyclical ups and downs.

How and why do these technology companies that survive the start-up perils do so well? For the simple reason that there is a great deal of perceived value in their products, and customers are willing to pay a great markup for that added value because even at that price they become more productive and more profitable themselves. A word-processing software package probably increases a writer's output by at least three times. For that, the writer is willing to pay $500 for some encoded electronic signals residing on a dollar's worth of floppy disk. The writer earns that $500 back in a month in increased output. The rest is profit.

The heyday of this entrepreneur/venture capital gavotte had its genesis in the mid-1970s when the potential of the microprocessor became apparent, at about the same time that the oil embargo dampened investors' enthusiasm for energy-related businesses. Clearly, the future was not in processing finite amounts of energy, but in processing infinite amounts of information.

Add to this the fact that there appeared a generation of young, ambitious managers in the established technical com-

Venture Capitalism

panies who had had the opportunity to deal with greater responsibilities at an earlier age than peers in older industries. That experience, plus the incentive of seeing others try and succeed, convinced many to begin finding their own way through the entrepreneurial swamp.

The primary thing that stood between the entrepreneurs and the venture backers in the mid-1970s was a high capital gains tax, which took forty-eight cents on every dollar of profit from a capital investment. That disincentive, on top of the risks involved and a recession at the time, virtually froze all start-up activity between the early and late 1970s.

In 1969, when the capital gains tax was 25 percent, over $170 million was invested in new companies. In 1972, the tax was raised to 48 percent, and by 1975, only $10 million in new money was committed to new ventures throughout the whole country. In 1969, there were nearly seven hundred public underwritings of small companies. In 1975, there were four.

Largely through the urging of the Palo Alto–based American Electronics Association, the capital gains tax was finally lowered to 28 percent in 1978. The effect was immediate. In 1977, $39 million was committed to venture capital firms to invest in new businesses. In 1978, it exploded up to $570 million. In 1980, the tax was lowered again, to 20 percent, and as a result, by 1982, $1.7 billion in new money was made available.[5] And much of that came from a heretofore untappable source: pension funds. A tacit approval from the U.S. Department of Labor in 1979 allowed a small portion of those enormous funds to be invested in venture capital partnerships. The inherent risk was no longer thought to violate the "prudent man" concept. In fact, the success of so many new companies made it downright imprudent not to be investing in them. And even a small percentage of a nearly $1 trillion-pool of funds represents a lot of dollars.

By the late 1970s, therefore, the dam had broken. The recent start-up frenzy began in earnest.

*   *   *

A number of people have made a passing comparison
between Silicon Valley and thirteenth-century Florence, the
birthplace of the Renaissance. The allusion is made to point
out that we too may possibly be on the verge of a similar
golden age, which likewise had its first flowering in a small,
out-of-the-way community.

The comparisons between that Northern Italian town and
this Northern California suburb merit more than a passing
glance. The developments that centered in thirteenth-century
Florence represent a good analogy for—and may help to
explain—the complex interplay between entrepreneurs and
venture capitalists going on in Silicon Valley today.

The technical and financial advances that took place in the
late Middle Ages in Northern Italy seem so primitive to us
now that we barely appreciate them for the ''leading-edge''
developments they really were at the time.

Two agricultural changes in particular in the eighth to tenth
centuries began the transformation. One was the discovery
that farmers could get a crop for two successive years from
the same plot. So they began to implement the ''three-field
system'' of farming, rather than the traditional two-field system.
Only one-third of a field had to be taken out of production
every year on a rotating basis to replenish itself, instead of
the customary one-half. At about the same time, a horse
collar was developed that rested on the horse's shoulder, not
around the animal's neck. What all this meant was that now
two-thirds, not one-half, of the arable land was available for
production in a given year, and horses could pull harder in
the harness with less chance of choking to death. With no
increase in resources, there was a significant increase in
productivity. Enough so that the surplus food and horsepower
meant that no longer did everyone, except the nobility, have
to labor in the fields.

One can almost hear the eleventh-century equivalent of a
social commentator wondering what would become of all the

Venture Capitalism

unemployed serfs. Could the economy ever expand enough to reemploy those now freed from agrarian servitude?

In fact, some of the labor was diverted into another basic activity: mining. And a new emphasis there prompted technical breakthroughs in that field: specifically, a water-driven bellows that helped automate the smelting process and, as a result, increased the supply of iron and gold.

An expanding food supply permitted a growing population, and a growing population demanded more efficient ways to grow more food.

Since not everyone was needed on the land anymore, some left and settled in towns to provide supplies—what are today called "productivity tools"—to those still on the estates. Iron was cast into plowshares. But people soon discovered that the expanding trade between craftsman and husbandman demanded a more sophisticated medium of trade than barter. So it was that in 1252, in the town of Florence, there appeared the first European coin minted in centuries. The florin, made from the newly available gold, allowed a standardized medium of exchange.

These symbiotic technical and financial advances created profound social changes. To keep those who remained still working on the land, overlords were forced to relax their feudal control. A new and prospering middle class appeared in towns throughout Northern Italy, and before long they were everywhere. Pope Boniface VII, in fact, placed them alongside fire, air, earth and water as "the fifth element."

It's not a coincidence that the town that minted the first coin of the age produced the first banks and banking families: Peruzzi, Frescobaldi, Medici. To serve the new demands of the new merchants, the banks developed new concepts: they took deposits, made change between currencies, and advanced credit in exchange for interest payments. And as the sophistication of the banks of Florence, Genoa and Venice grew, they opened branches in other European capitals. (London still has its Lombard Street.) The branches corresponded with

Venture Capitalism

their central bank and thereby made life easier for merchants who were increasingly trading beyond the olde town walls.

Merchants, too, became more sophisticated in their dealings, with the introduction of partnership agreements, marine insurance and double-entry bookkeeping, which allowed them a "real-time" look at the state of their business.

It's as if a dam had broken and the state-of-the-art technologies of the day—horseshoes, windmills, spinning wheels, canal locks, carpenter's planes, cranks, wheelbarrows—at last freed people from the daily struggle of eking out a precarious living, bound by hunger and the laws of the feudal order, barely managing a hardscrabble existence.

These advances in technology and finance played off on each other. A new society was created that no longer put a premium on stasis, but on dynamism. No longer was self-effacement a social virtue; composers, for example, began to sign their work.

Increasingly affluent merchants supported—demanded—brilliant works from brilliant artisans. It's staggering to realize that a provincial town the size of Florence could, over a period of three centuries, claim among its residents a pantheon that included Giotto, Ghiberti, Fra Angelico, Donatello, Cellini, Bramante, Uccello, Botticelli, Vasari, Raphael, Leonardo and Michelangelo.

All that work was possible not only because there was accumulated wealth to support it, but because the artist wasn't needed in the fields tending pigs, pulling stumps or draining swamps. And what a bitter irony that technology and art are so often seen as antithetical and at odds when, in fact, they support each other's existence. (Art underwrites technology if one accepts that real technical breakthroughs—the wheel, the plow, the microprocessor—are creative achievements of artistic magnitude.)

The advances fed on each other. Between 1200 and 1500, several technical refinements in shipbuilding and navigation—round-bottom hulls, the stern rudder, the astrolabe and the

Venture Capitalism

quadrant— were introduced that allowed the merchants of Venice and Genoa to trade farther and farther east. They brought back not only "spices"—seasoning, dye, medicine, jewels, linen, furs, carpets—but new mathematical concepts and medical treatments. The culture, the world outlook of all Europe, was changed profoundly.

My purpose here isn't to suggest a literal comparison between our two ages, but to indicate how a few basic, practical changes, interacting with each other, can go on to have long-term, widespread effect. We may well be going through a similar period right now, too close to it to see it fully and not farsighted enough to see where it's taking us.

The Renaissance didn't begin and end in Florence, and our next age didn't begin and won't end in Santa Clara. But both ages saw their roots take hold in those respective locations. That a single county in Northern California had some role in the development of the vacuum tube, the transistor, radar, the integrated circuit, the microprocessor and the venture capital partnership is remarkable, to say the least. (Santa Clara is still, however, waiting to recognize its Dante and Raphael, unless one considers news releases and video games to be high art.)

The venture capital deals can be as complex as shrewd investors, wary entrepreneurs and their respective counsels can possibly make them. But some general principles apply throughout which should serve as a background to the following profiles of three financiers.

Venture capital businesses generally fall into one of three categories: privately held partnerships, including family groups, with a general (managing) partner and limited (investing) partners; subsidiaries of mature corporations like GE and Xerox; and Small Business Investment Companies (SBICs), which are usually private firms regulated by the Small Business Administration in which private money can be leveraged up to four times by government guarantees.

Three states—New York, California and Massachusetts—account for over 60 percent of the venture capital pools in the United States, with California accounting for over 20 percent of the capital and the investing companies.

There's a popular misconception that venture capitalists are risk takers. While they are certainly more venturesome than bankers, they do not, by any means, see themselves as gamblers. They generally invest in only one or two percent of the hundreds of deals that come their way, and in recent years have come to expect about 30 percent of those to achieve above-average success, returning up to ten times the original investment over five years, the median showing a 25 percent compounded annual return over ten years.[6] Compared with the oil industry, where only one well in thirteen comes in, venture capitalists are not speculators at all.

Venture backers don't put all their money into a given start-up at one time, preferring to invest in stages or "rounds," so they can renegotiate the arrangement at various steps or cut their losses if need be. Definitions are loose, but the general arrangement flows something like this.

The initial investment is called "seed financing." It consists of several hundred thousand dollars to help the start-up team develop a prototype product, and get a workspace and a small staff to begin regularizing the operation.

The "first round" is invested while the company is still experiencing a "negative cash flow"—more money is going out than is coming in. It covers the completion of the product development cycle and the transition into the next phase of the company's growth: the making and shipping of large quantities of a product that up to now was handmade on a workbench.

By the time of second-round financing, the manufacturing line is coming up nicely and sample units are being evaluated in "beta" (trial customer) sites; the trade press has begun to

Venture Capitalism

cover the company, and customer leads are coming in based on that coverage and word of mouth in the industry.

At the time of the third round, the company is not only making money, it's keeping some of it; it has begun to show a profit. And by this time, the company's capital requirements are high and getting higher if it wants to be competitive, the financial press has begun to pay attention to the company (looking to the day it becomes publicly held), and the original venture capital source (an individual or a small group) has no doubt brought in other investors with more resources than it alone can provide. Those other investors also let the original source spread its risk.

Finally, if it has been successful up to now, the company "goes public"; that is, investment bankers take the company's stock, heretofore privately held, to the public market for trading. This last step gives the company access to the $1 trillion or more currently invested there by individuals or institutions.

The complete cycle can take as long as ten years or as little as two years or less. After closing out the cycle and watching the company go public, the venture capitalist is later able to reinvest some earnings and begin the process all over again.

(As with any economic activity, these cycles heat up and cool down, but the entrepreneur/venture capital phenomenon has followed this overall pattern.)

Following are profiles of three Silicon Valley financiers: one of the founders of the industry, Arthur Rock; a relative newcomer to the field, James Anderson; and Tom Volpe, whose firm, Hambrecht and Quist, takes companies to the public market to close out the entrepreneurial cycle.

9 | Arthur Rock

Arthur Rock was born in upstate New York, the son of a candystore proprietor. Though he was stricken with polio as a child, his favorite pastimes today are mountain trekking and skiing. Yet there is, he says, no psychological link between the venturesomeness of trekking in Nepal and venture investing in new enterprises. Rock's office is located in the Russ Building in San Francisco's Financial District. The building's elegant arched and polished marble foyer is from another, more stylish age, a distinct counterpoint to the glass-and-steel high-rises now sprouting throughout the city.

Rock's spartan suite is mecca to supplicant entrepreneurs all over the world, and contains framed prospectuses of some of the companies he has backed: Intel, Apple Computer and Teledyne. He was also the investor behind Scientific Data Systems, started by Max Palevsky in Los Angeles and later sold to Xerox for just under $1 billion. That was Rock's last

Arthur Rock

non–Silicon Valley venture, since he prefers to be physically close to his companies now.

Arthur Rock's backing alone helps assure a firm's success, because it generates attention and enthusiasm and attracts the best people. That initial prominence pays off, both for Rock and for the company.

He is a consummate pro. Prior to our conversation in his office, he had pulled together any reference material he would likely refer to and had it immediately at hand. He seems the epitome of cosmopolitan grace, and is certainly a man of considerable power and influence. Yet there is also about him a soft-spoken quality one would associate with a classical scholar.

He is modest about his success to the point of reticence. But his views on the qualities necessary for successful entrepreneurship—what he instinctively looks for in the people he backs, and how he is trying to promote an understanding of those qualities now by endowing a chair of study at Harvard—reveal a good deal about the man.

"When I went to Harvard I knew I wanted to go to Wall Street and make investments. But when you have nothing, what do you invest? I didn't have any capital, or very much capital.

"I went down to Wall Street in 1951 and started in the investment banking end of things. Gradually, I began putting together some deals. It was a different era; there was really no one putting together money to go into high technology. Even the term 'venture capital' was unknown until 1965. I think I was the first one to use it. I can't remember anyone using it before that."

Rock, as noted earlier, was the man who brought Fairchild Camera and Instrument together with the eight former Shockley employees in 1957. Rock subsequently found the Bay Area was where most of the new-technology deals were taking place, so he moved to San Francisco in 1961. He went into business, backing new businesses, with a former vice-president of the Kern County Land Company named Tommy Davis.

Davis and Rock existed for seven years, from 1961 to 1968. It was probably the first-ever limited venture capital partnership, at least as they are known today.

Then Davis went on to form another fund, and Rock did the same, called Arthur Rock and Associates, between 1969 and 1978. He had limited partners in that, too. "Now I'm on my own, doing it all myself. All the money I invest now is my own. I don't know of anyone else who has institutionalized themselves, if you will."

Arthur Rock has been quoted as saying that he doesn't have an understanding of technology. In fact, he says, he finds it would be an impediment to understanding the entrepreneurs he invests in. And for Rock, the investment is in the person, not the product. So what then are his criteria in selecting the people to back?

"The first thing I look for, and it takes a long time to find this out, is whether the guy is honest. I don't mean whether he's going to pick your pocket, or take your investment and go south. But I mean does he have the intellectual honesty to recognize his mistakes. Does he have the burning desire in the pit of his stomach to succeed. You know, you go out on the street and ask a hundred guys whether they'd like to be rich, and I'm sure all hundred would say yes. So it's got to take more than just a little desire. It's the sacrifices you have to make. And I don't mean sacrifices in terms of working twenty hours a day. But the ability to say no. That's a very tough thing for people to do, to say no."

And that means no to a lot of things, if necessary: no to people's favorite plans; no to others' and to one's own tempting but ill-conceived projects. To Rock, toughness doesn't mean ruthlessness in, say, firing someone coldly. But the

Arthur Rock

entrepreneur's toughness does have to extend to the entrepreneur himself or herself, in never losing sight of a goal, but knowing how to deal with every curve that comes along. And that ability requires balancing technical skill, diehard ambition and organizational finesse.

"I'm very selective in the projects I back. I don't do very many deals. But I've just been lucky in getting to know some very, very honorable people who really can make a go of it."

In a typical case, someone will call Rock, having been referred by a mutual friend or simply knowing of him by reputation, and offer to send in a business plan. "These days everyone has a business plan. Intel never had one. I asked them to put something in writing, and they put it in two pages." If Rock consents to read the plan and is interested, he begins a rather lengthy period of getting acquainted.

"I like to talk to these people over a long period of time . . . a month, two months, three months. Whenever they call up on the phone, I talk to them. Just trying to get a feel for what kind of people they are. Then we come to some agreement on terms and conditions.

"If the people have money, I like to see them invest it. You can't ask people to invest money that they don't have, but if people have money, I want them to invest it. I want them at risk."

Rock's philosophy, and it's deeply held, is that he has a lot more to offer a start-up businessman than money. And if the entrepreneur isn't interested in those other things which he has to offer, the feeling quickly becomes mutual. He prefers to become an active counselor: periodically attending staff meetings, becoming involved on the board of directors, recommending an accounting firm, law firm, public relations firm. In general, sharing his twenty-five-plus years of experience in nurturing some excellent companies from the ground up.

In fact, he sees one of his services as being a psychiatrist, of sorts.

Arthur Rock

"There was one company president who had a lot of problems running his business. He literally came up to my house and lay on my couch and started talking to me for about an hour or two once a week. He'd come up and have dinner or something and he'd just lie right down on the couch. That's carrying it a ways beyond what I normally do, but I am a sounding board."

Unlike some financiers who prefer an arm's-length relationship with their companies, Rock maintains an open-door policy: any officer of any of his companies can come in and talk with him. The typical reason, however, is to complain about the president. And either that person is right, or he's out of the company. Rock will not abide backbiting. If the manager has a grievance with the president, he takes it up with the president.

"I had one fellow come in and complain about his president. The president was having relations with his secretary on company time on a couch in his office, which I thought was guts. His personal life is his business, but what he does on company time on company property has to be my business. So we had to fire the president. And fire the guy that came in, too." But why fire the informer?

"Because this fellow was after the presidency, and he finally found a way to get it. He complained to me to get his president out so he could have the job. It wasn't that he thought it was for the good of the organization. But with this information, there wasn't much I could do with the president either, so they both lost their jobs."

Yet scalawags aside, Rock has invested in a remarkable string of world-class company builders: Noyce and Moore at Intel; Jobs and Wozniak at Apple; Max Palevsky at Scientific Data Systems; Henry Singleton of Teledyne. What is the common element he's found in all these men?

"They all had style. It wasn't always the same style, but having style is mandatory . . . the style of doing business, the style of operating. Charisma. But they all have style."

Arthur Rock

*   *   *

Arthur Rock's interest in selecting, nurturing, grooming and counseling entrepreneurs now goes far beyond finding business partners to invest in. He has recently become interested in identifying the qualities that are at the essence of entrepreneurship, to serve a much broader social need than his own investment portfolio. To this end, he and a colleague have endowed a chair at the Harvard Business School to promote the study of entrepreneurship.

"Fayez Sarofim and I met as students at the Harvard Business School and we have done a lot of business together, and I don't think . . . well, you can never tell what things would have been . . . but I doubt whether either of us could have been as successful without the other."

Sarofim is an investment counselor in Houston. He and his clients invested not only in the Davis and Rock partnership and in Arthur Rock and Associates, but individually with the respective companies as well.

"And, you know, that money was very valuable to me. And I made him a lot of money in return on his investments. As I say, I doubt whether either of us could have been as successful as we were without the other. So we just celebrated, and endowing that chair seemed like an appropriate thing to do."

An endowed chair at Harvard to study entrepreneurship is a remarkable idea considering that most American business schools have made their reputations in educating leaders of large corporations. There were always a few courses offered in the management of small businesses, but rarely was attention given to growing new enterprises from seedling to maturity within five years.

"Look at it this way. If you were an established business, and somebody came to you with an idea to make a new product in a new field, you'd say, 'Okay, what kind of an investment do I have to make? What kind of return can I expect? Over how long a period?' And the person with the new idea would say, 'Forget it; there's no way I can prove these things.' How is the management at General Electric or some other company going to accept this new idea? They build plants to build products where they have gone out and made a market survey. It takes a different spirit to pursue a new thing.

"A lot of people are going to have ideas that are just different and not acceptable. And they won't get done by the existing companies, because the existing companies can't do this. Most of the profits today are not made by companies that were making profits twenty or thirty years ago. The profit shift in American industry has been phenomenal. What we're trying to do is to help people learn how to accomplish this."

Because Rock has been at the center of Silicon Valley business development since the beginning, people often ask him what has been learned in this place that could be extracted and put to use to reproduce this phenomenon and stimulate this economic prosperity elsewhere.

"I've had delegations of people from various countries come in and ask me how do we do this in Ireland or Germany or Japan. You know, the first thing you need is an active stock market. Start from that end, because if there is no way to sell your securities in a successful venture, how are you going to get the money to put into a new venture? Most of these countries do not have that mechanism."

Two other components that promote entrepreneurship are necessary, as well. One is a business temperament that is at ease in high-risk situations, and that trait, at least up to now, has been uniquely American. The second is an infrastructure of support services that is geared to the fast-paced demands of start-ups: subcontractors, suppliers and vendors who not only have the needed goods but understand the pressures under which their customers live and are able to accommodate

Arthur Rock

them. An infrastructure like that does not suddenly appear overnight.

And for all its apparent negatives—shortage of skilled employees, expensive housing—Silicon Valley is still, for Rock, the place for a technology company to locate because of the rich concentration of other, supportive risk takers and the highly developed infrastructure.

Yet he sees that sooner or later, the boom here will become self-defeating. If spinning off and starting up continues *ad infinitum*, companies will no longer grow to the critical mass necessary to survive, much less succeed.

"I think this is a problem that people haven't really faced, and I am just beginning to think about it. Most of the established companies here that have sales of five hundred million dollars or more today are losing people. They're losing their good people, because some new company comes along and says, 'Why should you work for Intel or National Semiconductor or Hewlett-Packard for a thousand-share stock option? Join us, we're a new company, we'll give you five percent of the stock. If the company's successful, it will be worth a couple of million dollars.' That's a hard deal not to accept. And venture capitalists today are encouraging people to do this. What's going to happen to the established company that they've left, if all the good people leave? I don't know if there are enough good people to go around."

As one would hope of a man whose own success is assured, Rock is able to turn his attention to problems of a national scale, problems whose effect is on our society as a whole. And for him, the critical national problem as we enter an increasingly technical age is education. Excellence is, alas, not a term one hears much anymore in the context of public education.

"Everyone I know is sending their children to private schools. And I think it's terrible, but parents have no choice. I didn't go to a private elementary school. And no one I knew went to private schools. But it's happening now because the

Arthur Rock

parents think their children are not getting a good public education and they want them to succeed. I think as a result of this we're going to have an elite society, consisting of a very small percentage of those people who were fortunate enough to get a good education, and the rest are going to have mediocre educations."

The cure? "Those people who are gifted are going to have to be singled out. There were a lot of people getting no education, and the government decided, correctly so, that everyone is entitled to get an education. But with the limited funds, they have to teach to the average. We have to start teaching at the highest levels again, like we did years ago. It's a terrible problem we have in this country now."

One way he sees to address this, with schools strapped for funds, is for companies to fund more professorships and scholarships, or even set up their own degree-granting schools, as General Motors once did at its GM Institute in Detroit, educating engineers of all kinds.

There is one question I am most curious to ask Rock, the one question, it turns out, that every writer, neighbor, friend of his wants him to answer: How does he do it? What is it that he sees in one out of a thousand people that signals to him, 'This is the one to back'?

Rock's answer, every time, is the same: "I don't know." And he probably doesn't. Few artists, after all, have ever been able to explain how it is they do what they do.

What does this apparently shy man think of his life's work? Self-effacement doesn't describe his understatement when asked how it feels to have been involved in backing some of the most fundamental advances of our age: Fairchild's integrated circuit, Intel's microprocessor, Apple's personal computer. "Oh, pretty good."

There are probably a thousand entrepreneurs in the world who would dearly love an hour of Rock's time: his counsel alone, never mind the money. Yet the respect in which he is

Arthur Rock

held by others he in turn has for the people he has backed through the years.

"I'm the luckiest man in the world, that's all. I've just been very fortunate. You know, who ever gets to meet and know people who have done what these people I've invested in have done? There are maybe a hundred people like them in the world, and I've met ten of them. What are the odds of that?"

10 | James Anderson; Thomas Volpe

James Anderson represents a different type of venture capitalist, one with a technical, not a financial background. And one only recently involved in the activity. A graduate of Purdue, with an advanced degree in electrical engineering, he returned to his native California and went to work with Beckman Instruments, where he was involved designing such diverse products as aerospace equipment and cardiac pacemakers. After receiving his MBA degree from Stanford in 1977, he joined the computer systems division at Hewlett-Packard, working day-to-day in line operations—at the time, a not very prestigious career path for an MBA.

In the meantime, Steve Merrill, Jeff Pickard, and Chris Eyre, who had formed the nucleus of the Bank of America's in-house venture capital activity throughout the 1970s, decided to leave the bank and begin their own venture fund. Realizing it was better to retain some access to their collective skills, the B of A invested $40 million as sole limited partner in their new venture capital partnership.

James Anderson; Thomas Volpe

Merrill, Pickard and Eyre, with financial backgrounds, were looking for someone like Anderson with an operations background—someone who'd had hands-on experience with the computer industry on a daily basis—to round out their skills and bring technical expertise to their financially oriented decision-making process.

Among the enterprises that Merrill, Pickard and Eyre backed at Bank of America were Federal Express, Tymeshare, Advanced Micro Devices and Four-Phase Systems. Technology companies backed by Merrill, Pickard, Anderson and Eyre include Bridge Communications, Stratus Computer, Quantum, Priam and LSI Logic.

"My partners already had all the venture experience and judgment required. But there is a premium placed on being able to make accurate and rapid technical assessments in our business. I can interpret and assimilate the technical data that I hear from people more quickly and often with a different interpretation than my partners can. In addition, when I meet an entrepreneur, there's a more natural camaraderie between us . . . we're both out of operating backgrounds, and typically have shared common technology experiences."

Anderson's background gives him a better insight into the technical workings of the entrepreneur's proposed project. His on-line working experience better enables him to judge some of the nonfinancial aspects of the business plan, such as marketing strategies, product distribution, inventory control—"a lot of issues that most venture capitalists certainly could learn about but never bother to take the time to learn."

Still, he would never pass on a financing alone, without consulting with his partners, not only because of their financial partnership but also because of their experience in doing hundreds of deals.

Anderson sees another way his technical background has affected his outlook. "I think some venture capitalists would be just as happy investing in soap and cosmetics as they

James Anderson; Thomas Volpe

would in computers and disk drives. After all, we are judged solely on our rate of return, not the industries we support. But I'm emotionally married to high technology. I think it's great fun. I don't think of it as a way of making money . . . money is simply the way we keep score."

Because of his operations experience and training, there is a significant difference between Anderson's approach to judging a deal and that taken by an Arthur Rock. The latter, claiming little understanding of technology, gives primary consideration to the people involved in a deal. For Anderson, the product provides a mechanism to get to know the people.

"Arthur Rock's got a huge advantage over me, namely ten or fifteen years of venture experience I don't have. And I hope I get his kind of intuition, because I'm certainly going to lose any technical edge I've got now. If you're not involved in it on a day-to-day basis, you fall behind in technology because it moves so fast. Ultimately he's right; you bet on the people and you develop an intuition about situations. There's a set of vibrations that come off entrepreneurs and their approach. It's like body language. You're getting a lot by watching me beyond what my words convey. Well, Arthur is just taking that to an extreme. He knows what good entrepreneurs say and what money-makers should look and sound like. I don't have that depth of experience. I'm still tuning in. I do know what a good technology looks like, because I've been involved in those before when I was at H-P and Beckman. We're all trying to get at the same set of issues: Is this guy going to make money for me?"

However much he feels he still has to learn, Anderson thinks he's living out a dream: his own and that of many other technically oriented businesspeople.

"We're living through one of the most exciting periods in history . . . the intellectual level of these entrepreneurs, their energy, enthusiasm, dedication, combined with the kind of

James Anderson; Thomas Volpe

value they are able to create, is exciting. I'm thirty-three, and it's rare you get the opportunity at this age to play in that game. It really is fun, and people like Arthur Rock, who were fortunate enough to have witnessed the early years, have a much better perspective on it than I do. My concern, however, is that we are at the peak . . . that we're living through the heyday of an industry. Maybe it'll continue, I sure hope it gets better, but it could easily get worse. As the industry matures, much of the fun could be lost as we become institutionalized.''

And that concerns Anderson a great deal. The entire spin-off and start-up phenomenon may someday suffer from being too much of a good thing. As investors pour billions into venture funds, because of their recent track records, managers are under increasing pressure to get the money out, get it invested. As a result, it's feared many ill-thought investments may get made.

Already, some wonder if it's becoming a question of too much money chasing too few goods. After all, how many *really* new ideas come along in a given year? And then how many topflight managers are there to grow those ideas?

"I think a major issue facing the venture capital industry in the next five to ten years is going to be how to maintain the sense of closeness between the investors and the entrepreneurs in the face of large amounts of capital coming into funds. And you're already seeing mega-funds of several hundred million dollars, where you have ten or twenty partners, not just three or four. We're beginning to see the industry become institutionalized. Partners are experimenting with associates, and are spread so thinly they get further and further away from the entrepreneur himself, and even the investment decision-making process. Ultimately, I think this trend is dangerous. It will lead to investments being made in an imprudent manner.''

The days of individual patrons like Rock backing individuals he took the time to get to know—using inspired instinct as much as anything—may soon give way, Anderson fears, to legions of inexperienced deal backers making decisions, often too quickly, based on an MBA-style analysis, rather than an experienced perspective, of the entrepreneur's business plan.

The venture capital industry—made up in the beginning by men of entrepreneurial bent themselves—is no longer financially constrained, but it is experience constrained with what *Fortune* described in late 1982 as an "avalanche of newcomers."[7] And if many new to the business are attracted only by phenomenal successes like Apple (which paid back over $220 for every venture dollar invested), it's possible their enthusiasm may dim as quickly once those kinds of returns begin to stabilize. If they should pull out, what then becomes of all those new companies, not yet quite able to pay their own way but still viable businesses needing one or two more infusions of outside money? Could the funding fold up just short of self-sufficiency, leaving scores or hundreds of startups and their employees in the lurch?

"You see people throw money into an investment that is really ill-thought-out, and then when the business starts to crash, those same people don't have the experience, or the time or inclination, to help save it.

"It's fine for me to sit here and describe our portfolio of companies, some of which are going to make it and some of which aren't, and that's great. We're going to make money and so will a lot of other people and we're going to help America be more productive as a result. But the fact is that the vast majority of the people employed in small companies are never going to make anything close to the millions we all talk about. Those people are going to put in sixty-, seventy-, eighty-hour weeks chasing that dream, when the real reward may be the experience and not the value of the stock.

"The press is printing this hype about start-ups and venture capital, and we're feeding on it. I'm not going to sit each entrepreneur down and remind him of the risk he's taking by joining a start-up. But I can try to help him avoid mistakes

James Anderson; Thomas Volpe

I've seen elsewhere. What really saddens me is the tremendous number of entrepreneurial failures which result from mutual inexperience on the part of the investors and the entrepreneurs. These failures, done in innocence, are often the most tragic, but they're also the most difficult to avoid. It's tough to show someone who doesn't know something the significance of what it is he doesn't know.

"People look at the fact that at Apple at least a hundred employees made a million dollars as a result of the stock that they got, so that went pretty far down in the organization. But that is really an exception. Apple is one of those once-in-a-lifetime situations. But people still latch on to that and say that's what I'm going for."

In Anderson's experience, the senior people who start a company that goes on to do $30 million to $50 million a year in sales within three to five years can make $3 million to $5 million or more when the company's stock becomes publicly traded in the stock market. Middle managers can net from $500,000 to $750,000 in a similar deal—money they would never have seen if they had stayed on salary in a similar job at an established company.

"The problem is that for every one of those that makes it, there's a lot of other people out there that just put in a lot of time and get nowhere. . . . The sadder but wiser entrepreneur must either go back to a larger company, or go on to the next start-up in search of the financial brass ring."

Those hundreds of senior and midlevel managers, and thousands of production people, all have to dust off their résumés. Once again. But even in companies that survive, the growth is so quick that not everyone can handle it. Successfully overseeing a staff of two doesn't always equate to success managing fifty people eighteen months later.

"You cannot stand inexperience in a start-up. In an Intel or Hewlett-Packard if someone fails at a job, that's not the end of the world. They say, 'Let's find another job for this person and grow him . . . maybe we pushed a little too hard in that last one.' There's plenty of examples where large companies successfully grow managers from within. In a start-up, you can't afford to grow managers. If a guy doesn't work out, whether it's lack of background or whatever—maybe it's not even his fault, maybe he came in as a perfect candidate for the job, but the company grew too fast—it's not anybody's fault. But you have to either replace him, or bring someone in over him. There's hundreds of stories like that, that you see every day. I feel badly about that, yet all you can do really is, first, try to be conscientious in your hiring of people, and second, try to manage the people in the company to maximize their chances for success.

"A lot of people spend a lot of blood and sweat and never get compensated for it. I think that those are some of the unsung heroes of the entrepreneurial companies. Victims may be a better word—heroes in the sense that they tried and they failed, but victims in that they just got caught in situations that didn't work out. Often the problem is out of their hands. Bad luck or bad timing can cripple or kill a young company, and the entrepreneurs get nothing. Many can't go back to large organizations because they won't give up the challenges of a small company, so they often go on to other small companies and run the same risk all over again. You know, entrepreneurism is an infectious disease, and few of them ever return to corporate surroundings."

Yet clearly people continue to start companies, and seem to thrive on it. And Anderson himself stays in the business of financing those companies. "What salvages all of this is that the demand for good technical talent is far outstripping the supply. There are still so many opportunities that people do land on their feet. For now, that significantly softens the risk entrepreneurs are exposed to."

Lest someone thinks this issue is of consequence only to a handful of driven achievers and ambitious financiers, it's worth remembering that in all likelihood, it's these companies that will provide the majority of jobs of the early twenty-first century. Not only that, but portions of many retirement incomes are invested in the success of these venture funds. One way or another, a lot of people have a lot at stake in the success of the electron manipulators and their deal backers.

• • •

Although also a venture capital firm, it is as an investment bank that Hambrecht & Quist, Inc., headquartered in San Francisco's Financial District, has been called a "banker to the future." In that role, the company—like others in that field such as L.F. Rothschild, Unterberg, Towbin in New York; Alex, Brown & Sons, Baltimore; and Robertson, Colman & Stephens, San Francisco—serves to channel money from the public stock market to today's small, private technology firms that have survived the start-up throes and now aspire to be the Fortune 500 companies of the 1990s and beyond.

While venture capital helps take an idea and transform it into a marketable product and smoothly running organization, there comes a time in the life of that company when its capital needs—to grow as fast as the market it created, or just to keep up with the competition—become too great for any single investor, or group of investors, to satisfy.

At that point, the company sells part of the business to the public at large, thereby gaining access to the $1 trillion or so that is invested in the stock market. This step not only raises money for the company, but places public value on the stock and gives the venture investors, the entrepreneurs and the employees a chance to liquidate some of their shares.

Investment banks specialize in taking a company public, underwriting an initial (or subsequent) public stock offering. The custom today is for an investment bank to oversee the transaction over a period of several months, collect a fee, and move on. The philosophy at Hambrecht & Quist, however, goes back to the old European concept of a merchant bank that stays with a client from its early venture days, through the initial public offering, and on to any subsequent offerings.

This emphasis on a long-term relationship, plus management help, if necessary, and useful contacts with suppliers and customers, is appreciated in the fast-moving technology community, and is one reason the firm is so highly regarded, both by electron manipulators and in the financial community.

The company was started in 1968 by George Quist (who died in late 1982), former president of Bank of America's venture capital function, and William Hambrecht, who had been involved in underwriting small technology companies since long before they acquired their fashionable status.

In 1983, H&Q raised nearly $2.5 billion in the stock market for its clients—up from $700 million in 1982—although since it is privately held itself, it does not divulge its own earnings. That success wasn't always assured, however. In the depressed market of the mid-1970s, there was little interest in technology stocks. Hambrecht used to quip then that he was in the habit of sleeping only every other night. But the fact that the firm stuck with the industry almost exclusively, through thin as well as thick, is another reason it is so highly regarded today. It's known as a "pure play," a company deeply committed to its client base.

Tom Volpe is executive vice-president and chief operating officer at Hambrecht & Quist. At the time of our meeting, he oversaw that organization's investment banking function. He and his colleagues have about them an air of knowledgeable enthusiasm that, most certainly, is professionally necessary when working with a first-time company president about to embark on the biggest adventure in his firm's history.

Considering the amounts of money he deals with in the course of a year, Volpe's office is remarkably unassuming. In fact, it looked as though he was in transit—moving in or moving out—to judge from the boxes on the floor.

There is an air of nonchalance about the place regarding the work of raising millions of dollars. As we spoke, an associate came into Volpe's office to say that an initial public offering by a semiconductor company named LSI Logic that day had been successful, raising over $160 million from the stock market. "It makes my weekend," the associate said, waving off for the Friday afternoon rather as if he'd sunk a hole in one. To put $160 million in context, the gross national product of The Gambia in 1980, when it had a population of 600,000, was $150 million.[8]

The pros of going public, as mentioned, include the ability to raise money for the company and provide the initial investors an opportunity to cash out if they wish. The cons include the fact that the company is now indeed a public entity, which means it is subject to public disclosure requirements, regulations on its activity are set by the Securities and Exchange Commission, and there is pressure from the market to turn a profit every quarter when that may not be in the best long-term interests of the company.

Volpe explained the mechanics of going public. Before the company has selected which investment banker it wishes to deal with, the investment bank has to figure out, " 'Is this really a good company on a long-term basis? Is the management good?' The product life cycles in the technology world are getting shorter and shorter all the time, so what you are really betting on is the management's ability to anticipate and create new products in its marketplace. You make a determination—is it the product or the company that has been successful so far? Because a lot of things that go public are products, not real companies."

Examples of that would include software firms that have grown to quite good size based on the strength of a single product. While a company like that may offer other products, it is the milk from the "cash cow" that has kept the firm in business. The long-term (like maybe two years) risk in this situation is that once the market passes the product by, the company will be in real trouble.

*   *   *

But having decided that a particular company has long-term chances for success, the investment bankers, the company and their respective attorneys and accountants undertake a "due diligence" examination of the company's status and its prospects.

Based on that, a registration statement is filed with the SEC, and a "red herring" prospectus is circulated to prospective investors, so called because of the disclaimers printed in red ink. This tells as much as one could ever hope to learn about a company, its finances, its prospects, its management, its competition and "certain factors" that any prudent investor should know. The price per share is left blank; only a range of figures—between, say, $15 and $17 a share—is indicated.

"After about a month, the SEC comes back to us and says either, 'Okay, fine, you can go,' or 'We require certain additional disclosure.' Once satisfied, the SEC indicates that we can go ahead." In the meantime, H&Q has gotten indications of interest from institutional investors, retail investors and other investment bankers. With those other bankers, an underwriting group is formed with, say, H&Q acting as lead manager. This group literally underwrites the sale. That is, for a period of several hours or overnight, it buys and owns the stock issue prior to distributing the shares to the investors it represents. Hence the interest in spreading the risk among several underwriters. Among the companies which H&Q has taken public, such as Apple, Genentech, LSI Logic, Convergent Technologies, Apollo Computer and NBI, Inc., about 50 percent are in Northern California.

Volpe is in his mid-thirties. What background does it take for one of that age to be trafficking in billions of dollars a year? The desire and the training go back to school days, which for Volpe were at Harvard in the early 1970s—a time when going to business school was not a particularly popular thing to do.

James Anderson; Thomas Volpe

"Everybody thought I was absolutely crazy. 'How could anybody want to go to business school?' they asked. That was just the worst thing in the world. But I never aspired to be president of General Motors, and I never aspired to work in big business. It's great for what it is, but what it is and what I wanted were two different things.

"What interested me were people trying to do things a different way, and I was aware of it living in Boston then. I probably didn't call them entrepreneurs at the time, but I knew that there were these little companies out there then like Digital Equipment and Data General and Wang that were doing their own thing.

"I have always loved underdogs, whether it's a sports team or a company that's on the make. When you are constantly exposed to people that have come from nothing, then started a company and built something of lasting value, employing thousands of people, that has got to make an impression on you. Entrepreneurs are not born with silver spoons in their mouths. They're driven people who typically come from very humble backgrounds. And when you see as many of them as I do, you stand back and say, 'Boy, that's something!' "

After school, Volpe worked in the securities business in New York, which is where he first heard of Hambrecht & Quist. Though he was delighted to join the firm to open its New York office, he was reluctant at first to move to the Bay Area when the chance came up. Having made the move, however, it's safe to say he enjoys his work.

"I adore H&Q! I just love it! It's a fabulous place to work! But before you work here, you have got to have religion. You have got to believe that the technology sector is where the action is in this economy. That if this economy is going to survive and do well, it's got to come from here. This is where we really do add value in this country. I would say that virtually everyone who works here feels that technology is where it's at."

To that end, two of H&Q's seven venture capital funds are

composed entirely of investments made by executives and employees of the firm. What better way to encourage the faithful to keep the faith?

Yet in spite of the fact that Volpe clearly loves his work, he's frustrated by it as well. "When all is said and done, we are just intermediaries. The people who are really making it happen, doing what I consider to be important things from a macro economic and national-interest standpoint, are entrepreneurs. I am not the guy that is increasing the payroll from a hundred to three hundred people in twelve months. I take a great deal of pride in facilitating that guy's task, raising the money for him, but it's frustrating being an intermediary."

One of the reasons Volpe was reluctant to come West was the fact that "I have a lot of allegiances back East. I was born and raised there." And having been close both to Boston's Route 128 technology complex and Silicon Valley, he has done some comparing.

"In Silicon Valley it's always been socially acceptable to be an entrepreneur and have your own company. If you go to a cocktail party here and people ask, 'What do you do?' and you say, 'I've got my own company,' people here respect you for that. Back in Boston, it still means more to say, 'Well, I'm a young hard charger at Digital Equipment.' It's changing, but slowly.

"The culture out here is totally different. Being from Boston, I can appreciate it. Boston is a very exciting area, but it moves a little slower. You can also make the case that Silicon Valley moves too fast."

The question arises, to what extent can this dynamic economy be recreated on a national, perhaps global, scale? Will this half-county ultimately be the research and development lab of the next economic age? To Volpe, there are two issues involved if that is to be the case.

"A lot of people have lost their jobs in the last few years.

THE DEAL BACKERS 127

James Anderson; Thomas Volpe

In automobiles, in steel, in a lot of fields. Even as the economy rebounds, a lot of those people are not going back to their old jobs. And you can't just leave all those people not working. We have to get these masses of people who are skilled, but with the wrong skills now, retrained. It is possible to develop technology bases in Detroit, Chicago, Cleveland. But the first step is to retrain people, and not nearly enough has been done there, because people have the false hope that whenever the economy rebounds, they will go back to their old jobs. And they ain't never going back to those jobs.

"The second thing is that some mechanism has to be provided not only to retrain people, but to relocate them. You have a labor-constrained market in Silicon Valley and in Boston, and you have surplus labor in these other places. So labor retraining and labor mobility are the two key issues. And the mobility issue is one that can only be dealt with through government programs."

Clearly, something has happened in this valley. If it can only happen here, then this place is interesting but ultimately a dead end. But if lessons can be extracted and judiciously applied elsewhere, then—at its best—this place may point out what to aim for, and what to avoid, as we enter a new economic order.

Silicon Valley is a supposition, a hypothesis, a theorem. The conclusion is for others, elsewhere, to draw—either disregard, crib, or modify and localize.

4 | FROM WHENCE CAME?

11 | Catherine Gasich, Orchardist

How did the Santa Clara Valley ever become the Silicon Valley? How is one to account for all those miles of industrial parks sitting now where a few decades ago there were only prune orchards? What, in fact, was it like here not so long ago?

A number of factors—historical, academic, technical, financial and psychological—were crucial to this valley's metamorphosis. The story of that transformation is, in microcosm, the story of the genesis of the age of information processing.

In considering this location's transformation from saintly to sandy, it's worth pointing out that all the current dazzle—the successes of the Apples and the Seagates, the Hewlett-Packards and Intels—is a late-coming development. The Santa Clara Valley has seen its share of economic reversals: nearly two centuries' worth, interspersed with periodic bouts of bucolic quietude, repose, even somnambulism.

In vain, the mystical Spanish searched in Northern California for such mythical places as the Straits of Anian (the

Catherine Gasich, Orchardist

Pacific outlet of the fabled Northwest Passage), El Cibola, the Terrestrial Paradise, the Seven Cities, Gran Quivira and the greatest treasure of them all—El Dorado, the mountain of gold.

They found none of them. In the meantime, between 1769 and 1821, they colonized a small empire in California. Down the entire coast, a day's walk's distance on El Camino Real, they built a string of twenty-one missions, like Santa Clara de Asis; military installations called presidios, like the one at San Francisco; and civilian communities called pueblos, such as San Jose. And then proceeded to lose it all in short order when Mexico declared independence from Spain and took Alta California (from San Diego north) with it.

What followed was one of the most pastoral, arcadian societies in all history: "the splendid, idle '40s," or "the days of the dons." Caballeros and vaqueros lived to sing, dance, ride the range, hold rodeos, twirl *la riata* and court the señoritas on the balconies of the haciendas. There were no fences on the hillsides, and such was the hospitality of the Californio landowners then that there were no hotels down the entire length of their domain. Bed, board and even a horse were all offered gratis to the traveling stranger.

Yet in spite of the fact that California was about the most difficult place on earth to get to before highways and rail passage and air travel, the trickle of Yankee immigrants— spurred on by cries of Manifest Destiny—began to increase steadily with each year during the 1840s. And realizing it wouldn't be able to withstand the oncoming juggernaut, the weak Mexican government agreed to sign the Treaty of Guadalupe Hidalgo in 1846. Mexico got $15 million; the United States got California.

That was all legality. What really brought California to the attention of the rest of the United States was the discovery of a nugget of gold at John Sutter's mill near Sacramento in January of 1848.

"Great chances for scientific investment," said a San Fran-

Catherine Gasich, Orchardist

cisco newspaper in reporting the find. That would not be the last time that statement would be heard in Northern California. And when word of the find got beyond the state, the whole world rushed in, although because California was so hard to get to then, it did take the forty-niners a full year to arrive. If the mystical Spanish sought it in vain, it was the pragmatic Yankees who found it: El Dorado had been here all along. It was one of the few times in history when reality outstripped myth.

The impact of the find on those already in California, which is to say in the general area between Monterey Bay and Santa Clara, was immediate. One report has it a tent city of 4,000 disappeared overnight from San Jose. As the men of the South Bay Area headed a hundred and fifty miles northeast to the diggings, the women and children were left behind, and apparently they didn't have much political clout because soon thereafter the state capitol was moved from San Jose to the town of Vallejo.

The forty-niners set a pattern for the psychological development of California. It doesn't matter who or what you were back East, boy, you start all over again out there: clean slate, sky's the limit. In a very real way, the mentality that today creates new companies, new technologies and new industries on a regular basis—that pushes incessantly at the outermost limits in electronics, physics and biology—has its roots in the psychological makeup of these men who tossed everything over and trekked halfway around the world, hellbent to prove that their dreams could be realized.

It soon, however, became apparent to some of the early arrivals up at the diggings that not everyone who came was going to become a Croesus. A number of the argonauts, realizing that if they were ill-clothed and starving, others were as well, returned to the Bay Area and began to provide needed commodities like denim jeans and fresh produce. The valley at the south end of the Bay expanded its economic base from cattle—providing hides and tallow for the new industrial plants of New England—to farming and fruit growing.

But while the new orchards may have provided a needed product, the central role which the Santa Clara Valley had played in California's history up to then was over. With the gold rush and the later discovery of the Comstock silver lode in 1859, the focus of attention shifted north to the merchandising and transportation center of San Francisco and, later in the nineteenth century, south to Los Angeles. So after the gold rush, that collection of small communities at the southern end of the Bay, some of the oldest in California, seemed virtually to disappear from the mainstream of the state's history.

Once again, the Santa Clara Valley become arcadian, obscure and economically pinched, with the people there engaged in cultivating the small orchards, or else providing services to the growers.

Local promoters gave the place a series of names to attract the attention of outsiders: "the Heavenly Valley," and "the Valley of the Heart's Delight." But nothing seemed to lure people or industry, at least not in the numbers that were coming elsewhere. One has only to fly over Los Angeles and San Jose today to see which city got the lion's share of the growth in the past century. Even into the late 1960s, knowing the way to San Jose was considered something of a national joke.

In 1867, however, a man bought a farm midway down the San Francisco Peninsula. That unremarkable act set in motion the events that led directly to the creation of Silicon Valley here ninety years later.

Leland Stanford, one of the "Big Four" founders of the Southern Pacific Railroad, acquired part of the old Rancho San Francisquito located near a towering redwood known locally as El Palo Alto. His 7,200-acre farm might have remained just that—and, in time perhaps, become a housing development—except for the sudden death, by typhoid at age

Catherine Gasich, Orchardist

fifteen, of Leland Stanford, Jr., while on a vacation trip to Europe.

The old man has been variously described as slow of speech, ponderous, and, in the words of one contemporary, having "the ambition of an emperor and the spite of a peanut vendor." He came from very humble origins, but was no man of the people. He was one of the "railrogue foursome," an oligarchy perceived as utterly ruthless by many Californians.

But the father was profoundly stricken by the death of his boy. No rail line anywhere could ever bring him back again. Stanford underwent a significant change. It's said that after his son's death in 1884, he began to have dreams in which the boy appeared to him and urged him to begin putting some of his enormous wealth at the service of the community.

If there is a single moment that can be said to be the genesis of Silicon Valley, however unintentional, it probably occurred in Leland Stanford's dream-racked sleep.

The old man decided that on the third anniversary of his boy's death, he would dedicate a university on his farm to his son's memory. So in 1887, a cornerstone was laid.

And in October 1891, when Leland Stanford Junior University (which was not, as some thought, a junior university) opened its doors to its pioneer freshman class, the *San Jose Daily Mercury* saw fit to wax poetical: "As the Star of Empire westward takes its course, so also the Star of Education."[9]

But if the local paper took pride in the event, the university that aspired to become the "Harvard of the West" received little more than sneers back East. The idea of a great university out beyond the Rockies—located on a farm!—was ludicrous. About as ludicrous as the idea, two hundred years earlier, that the New World could ever have any school to match Oxford or the Sorbonne.

Leland Stanford spent his remaining years, until 1893, developing the school. With his death, his wife carried on the work. The old man, who might charitably have been described as a philistine in his lifetime, left an endowment of epic proportion to the West.

But university and pleasant surroundings notwithstanding, the Santa Clara Valley wasn't attracting people or industry on any epic scale, even into the twentieth century. Certainly not the way San Francisco and Los Angeles were doing. And it wasn't doing a very good job of attracting writers, either. No major work of American literature is set in the place, in spite of the fact that writers have commemorated the surrounding territory in all four directions.

John Steinbeck would write eloquently and passionately about the Salinas Valley and Monterey's Cannery Row to the south. Richard Henry Dana described the appalling plight of seamen on U.S. merchant ships in the Pacific in the early nineteenth century. Mark Twain and Bret Harte would use irony, sarcasm, melancholy and exaggeration to capture the life and hard times of the forty-niners in Calaveras County and Poker Flat to the east. And Dashiell Hammett would turn San Francisco's fog-shrouded hills into American myth with the adventures of Sam Spade.

Why industry avoided locating in the Santa Clara Valley is anyone's guess. I suspect that writers didn't use the pastoral locale as a setting because it lacked the fury, the brutality, the mystery and the conflict that are at the heart of literature.

Instead, this place seemed to alternate between splendid, golden, idyllic isolation and bouts of economic anguish. Sometimes both at the same time. Listen to the following account of daily life as it was lived in the Santa Clara Valley before 1945. It describes a world that, like Margaret Mitchell's Old South, is gone with the wind.

Catherine Gasich was, from 1935 until 1963, the postmaster of Cupertino. She was born in 1900, moved to the South Bay soon thereafter, and experienced the 1906 San Francisco earthquake, the Depression and the industrial explosion in the valley. Her home is located on a residential lane in the

Catherine Gasich, Orchardist

foothills of the Santa Cruz Mountains. Before the trees nearby grew in, she could see Coit Tower in San Francisco, fifty miles to the north, from her driveway.

"I was born in Healdsburg, California, and my folks came to San Francisco to live, but my father's health wasn't very good there because the weather was foggy and cold, and he had asthma. So we came down here to where it was drier. We were here at the time of the 1906 earthquake.

"I was sleeping in the bedroom and was knocked out of bed and came into the dining room and saw my father coming from the kitchen. We didn't have electricity in those days, and he was coming with a lighted lamp. It was five o'clock in the morning.

"I can still see him standing there holding that lamp like this in front of him, with his back in the corner of the room. And I looked the other way, and the floor was covered with dishes and rice and flour and broken eggs."

Soon after, the family moved to Cupertino, which, Mrs. Gasich says by way of correcting me and all other newcomers, is pronounced "Kewper-tino," not "Cooper-tino." Land was cheap in those days. And traded in some unconventional ways.

"An old Negro man names James Thomas had an acre of land. He was quite elderly and I don't know how my father got acquainted with him, but he, the old man, said that if you'll take care of me until I die, I'll give you the place. So they duly recorded that in San Jose at the Hall of Records. He used to have a bed that just swung like this, the spring was real rickety and he had an old patchwork quilt on it and the chickens used to fly up on his bed and lay eggs on it. I thought that was great. I remember when the old fellow died. The little old hearse drawn by a horse come and got him.

"My father was a cooper and he made barrels and tanks for the vineyards around. When you start a barrel, you've got to have some help. You've got this head, and you start with one stave and then you go around the head. I used to help him

Catherine Gasich, Orchardist

hold the staves until he came back to the first one, and then he'd put a hoop over it and tighten it.''

Fruit—prunes and apricots—were at the center of life in the Santa Clara Valley then.

"When I was seven, I started to cut apricots. We'd cut the fruit that had been picked, and it was taken to a dry yard and there were big trays made of redwood slats. And we'd lay the fruit on those, cut them in half, cup up, and then they were put into the sulphur house for about five hours. Then we'd take them out and put them in the sun in the dry yard. And that was the dried apricot that you get today.

"The apricot blossoms were pretty little applelike blossoms, pinkish in color. But the prunes, when they bloomed the whole valley was snow-white, and then this beautiful perfume would come towards you . . . it was just beautiful. We used to have Blossom Festival Days in the olden days, usually in mid-March, which were held in Saratoga. People from all around the Bay Area came to our valley to enjoy them.

"Land was cheap, but nobody had any money. There were some wealthy people, but most of the orchards were, I'd say, fifty, sixty or seventy acres. Some were ten acres or less.

"Raising fruit is a big gamble. There was the rain. That spoiled the blossoms. It washes off the pollen and the blossoms can't pollinate so then there's no fruit. Sometimes the frost comes along and kills everything.

"Most farmers sold their fruit to the cannery. We'd put it in the boxes and then the cannery came to get the boxes every day. Of course, you had nothing to say about what you were going to get for your fruit. The cannery set the price.

"Later on, they formed a prune and apricot growers' association here in this valley. Sunsweet, it was called. Many of the farmers joined that, and they got a little higher price.''

World War I, in far-off Europe, brought some changes to life in Cupertino. "The farmer was always an older person, and when his sons went into the service, we girls had to get up on a ladder and pick the fruit. We were called

Catherine Gasich, Orchardist

farmerettes. Before that, girls didn't do that. We'd cut 'cots and pick prunes, but we wouldn't get up on a ladder.''

Mrs. Gasich studied to be a teacher, but never pursued that career. Instead, she married and raised a family while her husband tended an eighteen-acre orchard. In the early 1930s, however, a career suddenly opened up for her.

"When Franklin Roosevelt came in, the postmaster here was a Republican. At that time, being postmaster was a political plum, and when a new president came in, he could appoint people he chose from his party. The postmaster was a man named Gilbert Aylesworth, and he was trying to raise his orphan brothers and sisters. He asked my husband, who was on the County Democratic Central Committee here in the valley, if he could help him. So my husband asked the committee if they would consider keeping Mr. Aylesworth on the job. They said no they couldn't, he had to go. Of course, then they didn't have any applicants, so they asked my husband, how about your wife, and he came home and asked me. About that time my older son was ready to go to Stanford and we thought maybe we could help him that way, because we just had a small acreage and things were kind of hard then, in the middle of the Depression.''

"At the Cupertino Crossroads, on Saratoga-Sunnyvale Highway and Stevens Creek Road, there was, in the early days, just a grocery store there and a blacksmith's shop and a church and a building which later became the Odd Fellows Hall.

"This grocery store was a general merchandise store. You could buy anything there—those blue chambray shirts . . . feed for your animals, except hay . . . various and sundry articles for the home or farm.

"The post office was in one corner of this store. When I was working at the post office in 1935, my clerks got twenty-five cents an hour, and I got the big sum of six hundred and

Catherine Gasich, Orchardist

seventy-eight dollars a year as postmaster, working from six in the morning until six at night.''

"Originally, this was called West Side because it's west of San Jose. And then some people petitioned that it be named Cupertino, because . . . see this creek that runs down here? When the padres came through here, in 1776, on March 26, they stopped on the creek and they named it after St. Joseph of Copertino in Italy. He had just been canonized. This was supposed to be called Copertino, but as time went by, the people didn't close the *o* and it became *u*. It's an Italian word, whereas mostly all the names in California are of Spanish origin.''

It took World War II to finally reintroduce the rest of the United States to the valley. "Many of the boys came from the East to go into the Pacific to fight. They'd come to San Francisco and maybe their ship wasn't ready to go out, and they'd have a few days and we'd hear about them. We'd put in our name, that we would like to entertain them. They'd come and visit with us, have a meal and stay overnight. And they just fell in love with this area.

"People here had just gone through the Depression. Some of them had gotten elderly, and they were willing to sell their land. Taxes were very high. The school tax alone was a hundred dollars an acre, and the farmers didn't make that much money. It was very hard for them to hang onto their land. So during and after the war, the subdivisions started coming in.

"I think it was a transition that just kind of happened. There was a little bit of electronics, mostly around Palo Alto. But it wasn't like it is today. Mostly radio and things like that. It wasn't all this stuff that we have now. It was much different. But little by little it just developed.''

"In '35, the population in Cupertino was about three thousand. Cupertino was six square miles, as far as the post office was

Catherine Gasich, Orchardist

concerned. After the war, things started booming, and they started building subdivisions and so forth and then cities got these growth pangs. They'd annex here and annex there, so that a good part of Cupertino was annexed by Sunnyvale, Santa Clara, San Jose . . . Saratoga took some. Now it's maybe two miles east and two miles west. It's all thickly populated now. There are hardly any orchards left in Cupertino.

"I was postmaster for twenty-eight years. I still had two years to go for my full retirement, but my husband was ill. He needed me more than the post office did, so I retired in 1963. Our orchard was taken from us by right of eminent domain, although we got paid for it. It was to build Interstate 280. Our house wasn't on the property we owned. We had built this house up here in 1945. It cost us thirteen thousand dollars to build. This is a well-built house, too. Right after we moved in, in 1945, people wanted to give us a hundred and fifty thousand dollars! I suppose today we could get, I don't know, maybe two hundred and fifty or three hundred thousand dollars. I have no idea."

By 1943, the development that Catherine Gasich talks about hadn't yet happened. And in that year, the *San Jose Mercury*, having had enough, decided once and for all that something had to be done about the low per capita income in the valley. The paper conducted a survey and found that few people outside the area even knew it existed. And those who did thought San Jose was a suburb of San Francisco, fifty miles away.

With funds from the City of San Jose and the Santa Clara County Chamber of Commerce, a national advertising campaign was put together, to promote growth and attract industry to the area. The Santa Clara Valley had little to offer in terms of natural resources, a large available labor pool or major transportation centers or markets, but the promoters decided to turn all the negatives to their advantage. During

Catherine Gasich, Orchardist

the war years, there was an emphasis in this country on decentralizing industry. The Santa Clara Valley was perfectly situated; it was miles from anything.

It was not a moment too soon for the promoters to become active. For while, during the war years, there was a big demand for fruit, that market declined after 1945. A drop in the price of fruit led to an economic slump in the area.

Yet in spite of all the setbacks and their apparent failure to be the magnet that San Francisco and Los Angeles had become, the communities of the Santa Clara Valley had unwittingly played their hand right.

The place had no outmoded industrial plant that needed to be cleared away or upgraded in the 1940s. It was a clean slate, desperate for business, with depressed land prices. It had an ideal climate, access to an incomparable variety of nearby ''getaway'' locations, and a university that was craving to establish its place in the sun. Just the attractions for companies that would depend on brainpower as the primary raw material.

And the upshot of it all was this: the industry that would soon transform Santa Clara was not back East after all, waiting to be lured out here. On the contrary, it was forming here all along—in a leaky attic laboratory at Stanford, in garages out behind the fruit trees, and in an almost unbelievable assemblage of young men, tinkering, wondering, investigating, and kicking around on dusty country lanes between Palo Alto and San Jose.

12 | Frederick Terman and
William Hewlett,
Patriarchs of Silicon Valley

       The transition from the Santa Clara Valley of prune blossoms and chambray shirts described by Catherine Gasich to the Silicon Valley of microprocessors and venture financiers is worth exploring. One individual more than any other accounts for that transformation, or at least the initiation of it. He was the pole around which so many of the formative elements—technical, academic, financial, psychological—were to gather and charge each other. He was a catalyst, an energizer, and apparently an inspiration. Today, the mention of his name is a litmus test in Silicon Valley. Those who recognize it knew the valley when. Many who are riding the crest of the wave left in his wake—one that seems to have no beginning, no end—have never heard of the man, much less his contribution.

His name was Frederick Terman. During his career, he was professor of electrical engineering at Stanford, but that no more describes him than to say that Walt Disney was an illustrator.

What Terman did in the 1930s and '40s, how he did it and the initial outcome of his activity are crucial to understanding the genesis of Silicon Valley. But to understand Fred Terman's role in creating the climate in which this center of high technology could happen in Northern California, it is necessary to go back even further in time than the 1930s.

California has a reputation for being a bit flaky, and a look at the early days of science and technology here will only add fuel to that supposition.

Consider the state's first eccentric millionaire, James Lick, one of those who tried his hand in the gold fields for a brief period, before returning to the Bay Area to buy some property in the Santa Clara Valley.

The richer his farm and millsite became, the more oddly he behaved. He wore ragged clothes, lived in a shack and slept in the frame of an old piano. It's said he couldn't pass up a bone in the roadway without examining it and, if it proved interesting, taking it home with him.

Mr. Lick is buried at the top of Mt. Hamilton, overlooking the Santa Clara Valley. He lies at the base of a giant telescope. That's his prerogative. It is also his endowment. Before his death in 1876, he created the Lick Observatory, now run by the University of California. The old coot also endowed the California Academy of Science and the Mechanics Institute Library.

From the earliest days, it was apparent there would be nothing conventional about the climate for science and technology here.

The early, fundamental discoveries in electrical and electronic engineering were made elsewhere, but the story of how many of these discoveries found practical application first in the Bay Area represents a string of successful experiments that goes beyond mere coincidence.

In 1888, a German physicist named Heinrich Hertz, working from the theories of a Scotsman named James Clerk Maxwell, conducted an experiment causing a spark to jump across a gap between two brass rods he had set up. That disturbance he created sent an invisible electromagnetic wave, a radio wave, through the air, causing another spark to jump across a similar gap—first with the receiving device located in the same room, then when it was located in another room with the connecting door closed.

In that first demonstration of wireless communication lay the genesis of the electronics revolution, when a high-frequency radio wave (today measured in a unit called a Hertz) could be sent and received through the air in some controllable, intelligent fashion.

Soon it was shown that such waves could be sent between buildings, between cities, for transmission of messages through space. And finally, in August of 1899, off the coast of San Francisco, from ship to shore. A lightship near the Golden Gate sent a dots-and-dashes signal to the Cliff House restaurant announcing, "The *Sherman* is sighted," bringing home the first troops from the Spanish-American War in the Pacific. This occurred a full month before Guglielmo Marconi conducted a similar experiment on the East Coast of the United States.

In January 1909, Charles David Herrold of San Jose, using an electric arc developed in 1902 by a Dane named Valdemar Poulsen that transmitted both voice and telegraph signals, became the first person to broadcast regularly scheduled radio programs for anyone "out there" who cared to listen to the sound of his voice. "Doc" Herrold had no call letters at first. With no competition, it was enough to say "This is San Jose" over the air.

In 1910, a young man named Earle Ennis, the proprietor of a store that sold wireless gear to ships, worked with a pilot whose name is lost to history and put a spark-gap transmitter on a biplane that was to circle Tanforan Race Track south of San Francisco. There was one major obstacle to overcome first: how do you ground a transmitter when it leaves the ground? Solution: you hook it to the engine. And that day, the Bay Area witnessed the first ever air-to-ground wireless transmission.

In 1909, a young man in Palo Alto named Cy Elwell took it upon himself to travel to Copenhagen to ask Valdemar Poulsen for the rights to manufacture his arc transmitter in the United States.

When Elwell returned to Palo Alto, he demonstrated one of the arcs he brought back with him. Dr. David Starr Jordan, president of Stanford, was so impressed by the demonstration that he invested $500 in Elwell and his new enterprise. Other faculty members and local businessmen came forth and did likewise. That may not have been the first, but it certainly wasn't the last time that the words "Stanford," "invest" and "new enterprise" would appear in the same sentence.

The newly formed Poulsen Wireless Telephone and Telegraph Company would soon change its name to the Federal Telephone and Telegraph Company. And in 1932, that company would be acquired by the International Telephone and Telegraph Company. But before that acquisition, Federal would send waves that would register not only across space, but across time as well. For in 1912, in a Federal Telegraph laboratory at 913 Emerson Street in Palo Alto, occurred one of those signal events after which nothing would ever be the same. In fact, that address has come to be called "the birthplace of electronics."

In 1911, thirty-eight-year-old Lee de Forest came to San Francisco from New York to supervise the installation, onto a U.S. Army transport ship, of a wireless telegraph set manufactured by his New York–based Radio Telephone Company. These units made use of de Forest's "audion," a glass-enclosed vacuum tube or electron tube he had invented back East in 1906.

Up to the time he came to California, the audion had been used only as a detector, or rectifier, to bring high-frequency radio signals, picked up by an antenna, down to a low enough frequency range to be audible to the human ear.

There are some who hold that de Forest was a born con man whose greatest skill was selling worthless stock in companies he would start and watch go bankrupt. Others hold that he was a genius inventor who was victimized by unscrupulous business partners. Whatever the case, de Forest's Radio Telephone Company suddenly went bankrupt while he was in California. Out of work and with nothing to return to in the East, de Forest asked Cy Elwell, then chief engineer at Federal, for a job with the company in Palo Alto.

Now, Federal had a big problem to solve at the time. With the growing demand for its telegraph service between California and Hawaii there was always a backlog, because receiving operators had to take the time to decode the Morse messages as they came in. Voice communication over that distance was impossible at the time, but if there were only some way to at least record the telegraph signals coming in, for later decoding by operators "off-line," the communications bottleneck would have been partially alleviated.

With two assistants, Charles Logwood, a radio amateur and experimenter, and Herbert Van Etten, a genius with circuitry, de Forest attacked the problem at the Federal facility at Channing Avenue and Emerson Street in Palo Alto.

At this time there was a primitive voice recording device available called a telegraphone, developed by Poulsen. However, electric dot-and-dash signals were too weak to be recorded on it. In order to amplify telegraph signals for such recording, so as to solve his new employer's immediate problem, de Forest fell back on his audion tube.

Finally, after a frustrating summer, a way was found to

wire three audions together in a manner which not only detected but amplified a signal. By 120 times!

And it amplified not just digital dots and dashes, but the analog sounds of a ticking watch, the footfall of a fly—and the sound of the human voice as well. Soon thereafter, in the spring of 1913, de Forest showed that his tube could not only detect and amplify but also generate continuous radio waves: the kind needed to transmit the human voice.

And because that "queer little bulb" could not only perform those functions but could oscillate and modulate and later switch and count, it would make possible such developments as long-distance telephony, sound-on-film recording, radio (in the sense we use the word today), televison, radar, microwave communication, electronic data processing and space travel.

De Forest soon left Palo Alto and returned East. In time he would invent a lamp for recording sound on film; he would establish a radio station; he would eventually sell all of his rights to the audion to American Telephone and Telegraph for less than $400,000 while being in and out of bitter squabbles with that company; he would see the Supreme Court twice uphold that he was indeed the first to generate signals with a vacuum tube; he would move to Hollywood and drive around there in a yellow Olds convertible; he would climb Mt. Whitney at age seventy; and he would die in 1961, at age eighty-seven, with nearly three hundred patents to his credit.

Other developments occurred here. On September 7, 1927, Philo T. Farnsworth, working at his lab at 202 Green Street in San Francisco, demonstrated the first successful all-electronic television transmission. Philo might be considered precocious—he was only twenty-one at the time of his demonstration— but he had been working on the project since he was thirteen.

The reception wasn't very good that night: it consisted of an out-of-focus triangle on a four-inch screen. But Farnsworth

Patriarchs of Silicon Valley

kept at it and in time perfected the system that twenty years later would begin to appear in homes everywhere.

There were many other young men here in those years, experimenting with the new technologies and forming new enterprises to market products based on those technologies. And there is a fine book that recounts in marvelous detail the personalities and achievements of those individuals. It's called *Electronics in the West: The First Fifty Years*. It was written by Jane Morgan and published in 1967 by the National Press in Palo Alto.

The book carries a preface by Fred Terman, who was, at the time, a provost of Stanford University. It also contains a photograph of Terman as a teenager, bent over a crystal set with Harvard and Stanford pennants hanging behind him on the wall of his room. A prophetic photograph, indeed. For in the fifty years between 1917, when the photo was taken, and 1967, when the book was published, Terman would cross-pollinate between those two schools and thereby earn for himself the title of progenitor of Silicon Valley.

Frederick Emmons Terman[10] was born in 1900 in a small town in Indiana. His father, Lewis, was a psychologist who believed in Rousseau's romantic notion of man as "noble savage." Lewis didn't send his son to school at first, but left him free to do as he pleased and learn as he would. Only when the boy was nine did the parents decide it was time for Fred to begin formal education. He finished primary school four years later.

In 1910, the family moved to Stanford, where Lewis worked on developing the Stanford-Binet Intelligence Quotient—the standard IQ test used in this country.

As a teenager, Fred was one of those numerous young "hams" in the Bay Area, experimenting with the new radio technology.

After graduating from Stanford in 1920 with a degree in

chemical engineering, he decided, like many Westerners then, to go East to gain legitimacy as a scholar. In 1924, Terman received his Ph.D. degree from the Massachusetts Institute of Technology, where he studied under one of the country's leading scientists, Vannevar Bush. He did so well in his studies at MIT that he was offered a teaching position there, to begin the following fall.

Before taking up his new post, however, he returned to Palo Alto to visit his family for the summer. While he was home, Terman developed a nearly fatal case of tuberculosis. To complicate matters, his appendix ruptured and he developed eye trouble. Because of the TB, he was forced to spend the next year in bed in Palo Alto with sandbags on his chest.

Terman somehow managed to survive his ordeal. And in 1925 he was able to take a half-time position at Stanford, teaching electrical engineering. Though still bedridden twenty-two hours a day, he was able to get up long enough to handle his two-hour course load.

While suffering his enforced quietude, he became more and more interested in radio technology, as it was called then, and several years later, in 1932, he wrote a book called *Radio Engineering*, one of the classic texts on the subject. Terman designed the book not only for use in the classroom but for the industrial reader as well. His interest in both domains would remain his lifelong hallmark.

By 1937, he was a full professor at Stanford and head of the Department of Electrical Engineering. He was also beginning to get very tired of the brain drain he saw around him every year: the migration of his students eastward to find job opportunities that didn't exist in California then.

The problem bothered him so much that he began to encourage some of his students, who worked in Stanford's leaky attic radio laboratory, to consider starting up businesses of their own on the San Francisco Peninsula. Two students in particular had often talked about doing just such a thing. But when he graduated from Stanford in 1934, David Packard, like so many others, went East, to join General Electric in upstate New York. His fellow student, William Hewlett, remained at Stanford to do graduate work.

"A new idea in electronics turned up," Terman would later recount. "I told Bill, 'It looks to me as if you could use this to make an instrument. It would be a lot simpler and cheaper than anything on the market. But you'll have to solve a couple of problems to make it function.' Bill came up with an absolutely perfect solution in a brilliant way."

The device was an audio oscillator, a unit that would generate signals of varying frequencies.

"Money was a problem," Terman said, "but by great effort and with a bit of luck I was able to get some money together for the project.

"We wanted Packard back because he had just the quality I preferred. So we spent five hundred dollars for materials and five hundred dollars for Packard's salary. He took a leave of absence from his job at GE [paying one hundred and ten dollars a month] to come back here [for fifty dollars a month]."

With Terman's continued encouragement, the two young men started their fledgling business in Packard's garage. The first customer for Hewlett and Packard's audio oscillator was the Walt Disney Studio, which used it for recording the sound track of *Fantasia*. After that, Terman could tell how business was going for the two without even asking. "If the car was in the garage, there was no backlog. But if the car was parked in the driveway, business was good."

With the outbreak of World War II, Terman's teaching career came to an end. His former professor Vannevar Bush had been tapped by President Roosevelt to organize the efforts of the academic and scientific community for the country's defense. Terman was asked to return to Cambridge to head a radio research laboratory that would develop countermeasures against enemy radar. Starting from scratch, Terman soon organized over eight hundred people, most of them under thirty, for work so secret he couldn't even discuss it with them at first.

They began their efforts in early 1942, and by late 1943 all Allied bombers were using their jammers and chaff (narrow strips of aluminum) to throw off Axis radar. One estimate holds that seven hundred Allied aircraft were saved as a result of this effort.

"The war made it obvious to me," Terman said later, "that science and technology are more important to national defense than masses of men. The war also showed how essential the electron was to our type of civilization. Our method of fighting the war was to an amazing degree built around, and determined by, electronic devices."

Though his teaching days were over, Terman's education was not finished. "I learned a lot about Harvard—how it ran and why it ran. I found out that it is essentially run by the president, treasurer and five fellows, who between them constitute the 'corporation.' These people were approachable.

"My neighbor in Cambridge, the treasurer, would work every Sunday in his backyard. He was an avid gardener. I'd stand around behind him, following him as he worked.

"I asked him what he thought would happen after the war. It seemed to me there'd be a new wave of government research support—the scientific war effort had been so very successful. Well, he told me that the Harvard corporation had been thinking about this, too. They were wondering how Harvard would handle the research. His views, as it happened, reinforced my views.

"Harvard wasn't going to do big project research with lots of people. They'd have small projects, each one-man size.

"The idea of peacetime support of university research by the government was new. The U.S. had never before World War II provided any substantial support for basic research in universities except in the field of agriculture."

When he returned to California at the end of the war, Terman had the challenge of putting his ideas into some plan of action.

He was instrumental in setting up the ground rules for postwar "sponsored research" at Stanford.

Patriarchs of Silicon Valley

The work would need to be of significant academic value and require a minimum of administrative supervision. Furthermore, it would be geared to the interests and expertise of individual faculty members who would do the work themselves, in exchange for reduced class loads. Graduate students could be recruited to help, would be paid for their efforts, and could use the findings in their dissertations.

"Stanford emerged from World War II as an underprivileged institution. It had not been significantly involved in any of the exciting engineering and scientific activities associated with the war. When I returned in January 1946 to take over as Dean of Engineering, I started from scratch, without forward momentum. But we had a chance to achieve a stronger national position.

"The universities came into the picture only because it has been found that they are the most effective places that we have to carry on fundamental research. Private philanthropy would have been inadequate. Industrial needs must be profit-oriented. States could not assume this national responsibility. A natural result of this situation is that government money tends to flow to universities, and to those parts of universities, that possess strength. They wanted first-class members to carry on their research.

"The West has long dreamed of an indigenous industry of sufficient magnitude to balance its agricultural resources. The war advanced those hopes and brought to the West the beginnings of a great new era of industrialization. If Western industries and Western industrialists are to serve their own enlightened and long-range interests effectively, they must cooperate with western universities and, wherever possible, strengthen them by financial and other assistance."

Terman took to the hustings, promoting the school to industry and vice versa. He spent numerous evenings speaking at community and businessmen's dinners, and by the 1950s served on the boards of several new Peninsula enterprises. He set a pattern that continues to exist. Today, over 60 percent of Stanford's engineering faculty consults with government or industry.

The symbiotic relationship between Stanford and local businesses worked to the benefit of both parties. By the mid-1950s, corporate gifts to Stanford amounted to $500,000 a year. Ten years later, the level of contribution was up to $2 million a year. And by the mid-1970s, it approached $7 million. Stanford, in turn, worked to the benefit of the local business environment by educating engineers and businessmen in droves.

One of the first things that Stanford did after the war was to begin leasing sections of Leland's farm to emerging technology companies like Hewlett-Packard. The Stanford Industrial Park became the first major park of its kind in the country. Today it covers over eight hundred acres, and income from some ninety tenant leases supports the school.

Terman's goal of creating a center of high technology—"a community of technical scholars"—was, in his words, a throwback to the great medieval universities at Oxford, Cambridge, Heidelberg, Bologna, Göttingen and Paris.

He succeeded beyond his wildest dreams. Or maybe it wasn't beyond his plans. Maybe he foresaw it all along. Time and again, when people here are asked what there is in Silicon Valley that can be transplanted elsewhere to recreate the phenomenon, the response is: Get another Fred Terman.

Among those who can most keenly appreciate Terman's contributions is William Hewlett, a founder and now vice-chairman of the board of Hewlett-Packard, today a company doing almost $5 billion in business around the world, with its headquarters still in Palo Alto. By virtue of its size, its longevity and its reputation as a manufacturer and employer, H-P is generally regarded as the premier Silicon Valley company.

"I think Terman's impact was in several areas. In the first place, he wrote the definitive book on radio engineering. He pulled a lot of information together and did a fine job of

Patriarchs of Silicon Valley

collating it and presenting it in a unified fashion. The book had a profound effect on the upcoming generation of what were to become electronic engineers, though the name hadn't been invented by then."

Terman's second contribution, according to Hewlett, had to do with his concept that there could be a synergistic relationship between applied technology, as it was then called, and the university. "I'm pretty sure that he was responsible for the concept of the industrial park which is on some of the acres around the university where H-P and a number of other high-tech organizations set up facilities."

Hewlett recalls that Terman was an incredibly thorough man. "He didn't have a narrow mind. He was interested in many things. I remember one day he said, 'You know, the Finns have been throwing the javelin much farther than anyone else, and I took a look at this and I realized they had developed a technique and it won't be long before other people come along and develop that technique.' And sure enough, other countries came along and did just that.

"When you started a discussion with Fred, you couldn't get him off the subject. He had something to say and he was going to say it. He was a very clear thinker, and presented his thoughts very well."

The radio laboratory at Stanford was upstairs in the engineering building. "The school was pretty poor in those days, and when it rained, it leaked in, so they had like a sandbox with tarpaper to catch the water, and it collected there. According to Terman, one day I put some goldfish in there. Whether it's true or not, it's a good story. Because it expresses the spirit of that laboratory. It wasn't all work. We had a lot of fun, too."

Aside from the work Terman led at Harvard, developing defensive measures for aircraft, two technological developments were made in the Bay Area in the 1930s that profoundly affected the Allied victory in World War II.

Across the Bay from Stanford, at the University of California in Berkeley, physicist Dr. Ernest O. Lawrence was developing the cyclotron, a circular device that would accelerate a stream of protons until they were moving fast enough to bombard and break open the nucleus of a targeted atom.

In 1931, Lawrence was conducting his work using a magnet only four inches in diameter. If he was going to take his research any further, however, he would need a larger magnet. But it was the Depression and money for research projects was hard to come by.

Then one day over lunch at the Faculty Club at Berkeley, Lawrence mentioned his problem to Dr. Leonard Fuller, head of the university's electrical engineering department and, coincidentally, an executive vice-president of Federal Telegraph. After listening to Lawrence, Fuller offered him an unused Federal electromagnet with poles forty-five inches in diameter, ten times the size of Lawrence's existing device.

The eighty-five ton object was brought across the Bay to Berkeley, and with it, Lawrence carried on the research that, in 1939, won him the Nobel Prize. With the cyclotron, Lawrence and his co-workers were not only able to smash atoms but to "create" atoms like berkelium and californium, which didn't exist in the natural world.

(Dr. John Lawrence, at the University of California's medical school, would later make use of his brother's device in the treatment of cancer cells.)

Among Lawrence's colleagues at Berkeley then was Dr. Robert Oppenheimer, who would later be in charge of the Los Alamos Laboratory and figure prominently in developing the atomic bomb.

But it's worth remembering that the purpose of the early atomic research was not to explore the potential of nuclear fission as a weapon. In 1931, Lawrence and his colleagues "merely" set out to see how the material world was put together.

*   *   *

Patriarchs of Silicon Valley

The second significant prewar development here grew out of an interplay between two brothers, Russell and Sigurd Varian.

Russell didn't show much promise at first as he and his brother were growing up together in Palo Alto. In fact, he appeared to be a slow learner, not graduating from high school until age twenty-one. When he was admitted to Stanford, he decided to major in physics because he thought it would require less reading than any other field.

As a young adult, Sigurd's interest took off in another direction from his brother's. He became, in the years just after World War I, one of those daring young men in their flying machines whose self-confidence and élan took them farther than their flying machines were ever designed to go, crashing numerous times at night, in fog and during storms.

In the mid-1930s, Rus and Sig got together as adults to form a laboratory, as they had often talked about doing when they were tinkering together as children. They teamed up with a former Stanford classmate of Rus's named William Hansen, now teaching physics there, and began to work on the development of a device that would solve the pilot's perennial problem: flying blind.

The three men set to work on a tube to generate waves of high enough frequency to sweep the sky at some distance, and return a signal to the sender when there was an object "out there" in the way.

The men needed backing, however, and once again, Stanford came to the assistance of enterprising engineers.

The Varian brothers were made "research associates" at the school. They could draw no salary, but they would have access to the school's physics laboratory. The school also provided them with $100 for expenses, in exchange for a share in any return that might come from the work.

The tube they created was given the name "klystron." It could generate and amplify microwaves, as de Forest's tube generated and amplified lower-frequency radio waves. And while these microwaves are invisible to the eye, the waves created by the klystron in the macro world would become very visible in the eyes of history. For by now, the Battle of Britain had been joined in nighttime skies over a darkling plain, and the English-speaking world held its breath.

The klystron tube, weighing less than six pounds, was a negligible addition to the payload of Royal Air Force planes. Sig Varian's long-held wish had come true: pilots had on-board radar; they could see in the dark. It's said that Hermann Goering thought it was unfair when he was told that British Hurricanes and Spitfires could hover at high altitudes in the dark nighttime sky, even with cloud cover, and take their toll of his unsuspecting Heinkels and Junkers.

The technology of World War II—with electronic and atomic roots in the Bay Area—offers perhaps the ultimate study in paradox. On one hand, the use of radar to defend the homeland; on the other, the use of atomic fission to vaporize civilian populations.

In 1948, the brothers Varian, William Hansen and some others formed Varian Associates. In 1953, that company was the first firm to locate in the Stanford Industrial Park, where it continued to develop the klystron for use in military and civilian communications applications. And Stanford University would, in time, realize over $2 million in royalties from its original $100 investment.

The end of World War II brought a period of self-doubt to the fledgling electronics-based companies in the Bay Area and elsewhere. Would there be anyone to use their products when the prime customer—the Pentagon—wound down its efforts?

A Russian émigré named Alexander M. Poniatoff faced that dilemma. Because he had wanted the hallmark of his wartime motor manufacturing company to be excellence, Poniatoff added "ex" to his initials when he named his firm Ampex. (If he could have been assured of a royalty from every technology firm that would later crib his suffix-x example he probably could have retired comfortably on the spot.)

Patriarchs of Silicon Valley

Shortly after the war, a business associate told Poniatoff about a demonstration he had recently seen of a device captured from the Germans that recorded sound on magnetic tape, not, as customary at the time, on disk.

The "tape recorder" was a significant advance over existing disk recording technology on two counts: the sound was much more lifelike, and tape, unlike disk, was editable.

Here, Poniatoff thought, was just the potential business he was looking for to keep his own enterprise up and running. So in 1947, Ampex introduced its Model 200, which, the company claims, was the first practical professional magnetic audio recorder to be offered commercially.

What assured the product's success was the decision by Bing Crosby to purchase twenty units so that he could prerecord his popular radio program, relieving him of the pressure of performing live and giving his producer the chance to edit out any fluffs.

Ampex had made the transition to the peacetime economy. More significantly, it showed there was a tremendous commercial, and maybe even consumer, market at last ready for the new technologies.

To conclude this account of the genesis of Silicon Valley, there is one last aspect that merits consideration. But this has less to do with Santa Clara County than with the state of the State of California in those years.

Prior to World War II, California surprisingly was not a major economic center. The only significant businesses in the state then were agriculture, tourism and real estate development. But all that changed with the bombing of Pearl Harbor. Because of the national mobilization, particularly for the war in the Pacific, California prospered enormously. It had no outmoded production facilities to tear down; it wasn't tied to obsolete technologies or ways of doing things; and it was no sooner building its new facilities—for constructing aircraft in Los Angeles and ships in the Bay Area—than the products of those factories were sold to the government.

California hitched its wagon to a Pentagon-shaped star during World War II, and as a result it reaped immediate, large-scale rewards in defense and military-related work. Men who had only two years before been begging for a day's work, scorned as "Okies" and "Arkies," were now very much in demand.

In 1937, the entire aircraft industry employed 30,000 people. In 1940, President Roosevelt called for the production of 50,000 planes a year, and small firms in Southern California with names like Lockheed, Douglas, Northrop and North American were suddenly finding their services very much in demand. In 1937, Lockheed had about 1,400 employees and sales of $5 million. Within only a few years, the company employed 53,000 and had sales of $145 million. By 1944, over two million people worked in the industry, and the Douglas Corporation alone produced $1 billion worth of planes that year.

Northern California, too, suddenly found itself in the midst of unprecedented manufacturing activity, building ships, not aircraft. Of the $5 billion worth of vessels contracted for by the government during the war, $3 billion worth were built in the Bay Area.

With this level of activity, the state's industrial workforce tripled within only a few years, from 380,000 to over 1.1 million. California ceased to be strictly an agricultural state, but the state's farmers, too, saw their markets and output expand by a factor of three between 1939 and 1944.

The newly created wealth reached through the corporations and the corporate farms—2 percent of the farms in the state occupied over 60 percent of the total acreage—to reach individuals. During the war years, the general income in the state rose from $5 billion to $13 billion, and average individual wages increased from about $1,930 to $2,950 per year.[11]

It's hard to believe that only a few years before, in the late 1930s, California was close to having a revolution on its hands, as displaced Dust Bowl residents stood hat in hand begging for work, and starving families saw milk poured in ditches to keep the price up.

The boom times that hit California continued after the war. The pace only quickened with the Korean Conflict, the "missile gap" and the space program. In the years after the war, something like one out of every three manufacturing jobs in the state would owe its existence to a defense contract. Fifty percent of all National Aeronautics and Space Administration contracts were let in California. Nearly 60 percent of all manufacturing jobs in Los Angeles–Long Beach, and 70 percent of San Diego's workforce, dealt with defense-related work.

Between 1955 and 1965, the number of manufacturing jobs throughout the United States would increase by 6.5 percent. In California, in those years, the number of those positions would increase by 25 percent. At the peak of that postwar prosperity, the state saw 250 acres *a day* become urbanized to house the people in all the new jobs.

All of this had a tremendous psychological effect on a generation of young men in this country who came home from the war having spent their entire lives up to then watching dreams be deferred. Born, weaned and reared during the Depression, they then spent the best years of their lives in trenches, submarines and hospitals, waiting for the chance to make their own destinies. When they returned victorious from the war, they suddenly found they lived in the richest, most powerful country on earth. They had waited long enough. They wanted the chance—at long last—to get on with their lives.

A million and a half men passed through the Port of San Francisco on their way to and from the Pacific Theater. Some later settled into the coffee houses of San Francisco's North Beach and—whether beaten down by their experiences, or made beatific by them—became the Beats. Some of them, as Catherine Gasich said of her guests, had come down the

Peninsula for an overnight visit during the war years and fallen in love with the area. A number of those returning veterans, exposed to the new technologies like radar, microwave communications and vacuum tube technology, were craving the chance to make a living from the skills they had picked up in the services. And hearing of the new technology companies here, like Hewlett-Packard, Ampex and Varian Associates, they headed south down the Peninsula after the war. They bought lots in the subdivisions that had only recently been prune yards, and they transformed Santa Clara from rural to suburban.

Yet one feature would always distinguish Santa Clara County communities from every other suburban area sprouting up across the country after the war: these were suburbs without an "urb." San Francisco, fifty miles north, was too far away to be considered adjoining. And San Jose, in the 1950s, was not yet a major metropolitan area. If the emerging communities in the South Bay had any center at all, it wasn't a city or a market, but a university. Seldom since the Middle Ages had that been the case.

When William Shockley located his new semiconductor laboratory here in 1955, the metamorphosis was complete.

Consider a spring morning back in 1913 in Palo Alto. On Emerson Street, a forty-year-old Lee de Forest and his colleagues are working to find new applications for the vacuum tube. On Bryant Street, the teenaged Varian brothers are, as usual for them, tinkering away. Nearby, a thirteen-year-old Fred Terman is sending and receiving messages on his crystal set. And on Waverly Street, a three-year-old William Shockley moves into a new home as his father takes up a teaching position at Stanford. And all the while, the air is fragrant with the smell of orchards in bloom.

How does one account for this assemblage of talent, congregating near a deceased railroad baron's old farm? Is it only coincidence, or do some places bristle with higher-than-average energy fields?

13 | George Morrow: Personalizing the Computer

     A transformation took place here between the days when heavily engineering-oriented companies like Shockley Semiconductor were typical of the place, and today, when consumer-oriented companies like Apple are more representative, in the public's imagination, of Silicon Valley. In fact, it was that transformation, represented by the personal computer, that brought Silicon Valley to the world's attention and stamped the place with its popular identity. Curiously, much of the impetus for that change came not from Santa Clara but from East Bay communities like Berkeley and San Leandro.

One of the catalysts of that change is a singularly colorful iconoclast, possessed of a bald top, squinting smile, and rolled-up jeans and running shoes, who exhibits the inscrutable countenance of a kung fu master and the habit of a jogging gardener.

''Unconventional'' would be a good, if understated, word to introduce George Morrow. And though he never set up shop in Santa Clara County, he has observed at close hand and been a part of the transformation which took place here, and throughout the entire microelectronics industry, between the early 1960s and today.

In common with a lot of unconventionally bright people, Morrow was a high school dropout. And until he returned to the academic world at age twenty-seven, he romped through a life filled with activities that never manage to show up on a professional's résumé. He bummed around the country for two years; served in the armed forces in the Far East, and remained there for several years after his discharge; was a mechanic, cat skinner (someone who operates Caterpillar construction equipment) and construction worker back in this country; and got a lot of experience as a short-order cook. In all, he figures he had about 150 jobs during the ten-year hiatus in his formal education.

But no matter what the work entailed, he found that sooner or later all the jobs he had got to be like making deviled-egg sandwiches as a cook: tedious and repetitious. And once again, it would be time to move on. Eventually, he saw that there was no possibility of getting any kind of challenging work with the skill set that he had. And, too, by his mid-twenties, he had come to terms with his own private battles.

"I was at war with the whole process of having certain kinds of rules, certain kinds of rights or wrongs, imposed on me as a young person. I finally decided that maybe instead of trying to wholesale fight the thing, I would accept the positions that I didn't have any argument with—aside from the fact they didn't originate with me—and only concentrate my battles on the parts I wasn't comfortable with."

At twenty-seven he enrolled at Foothill College in Los Altos, majoring in physics, and took a part-time job as a technician at Shockley Semiconductor. It was the early 1960s, shortly after the eight Shockley employees had left to form Fairchild.

It happened that two managers at Shockley—one in re-

Personalizing the Computer

search and one in production—each wanted Morrow in his group. The issue reached an impasse, and, as was the custom there, the matter was brought to Shockley himself for arbitration.

While sitting with the two managers in Shockley's office, waiting for his decision, Morrow claims to have overheard a phone call that gave him an insight into the heart of the entire matter of Silicon Valley spin-off and start-up activity. Several Fairchild employees had recently defected from that company to form another offshoot called Signetics, and the management at Fairchild was apparently thinking of suing the renegades.

"I'm sitting there in the presence of the great man, waiting for him to decide on my issue, and he is saying on the phone to someone, 'You tell those young turks at Fairchild that if they sue Signetics, I'm going to sue Fairchild.'

"I think he set the stage for all the continued defections in the Valley. Because everyone was scared to death that if anyone started lawsuits, Shockley would start lawsuits over the defection of the Fairchild people. And if he'd sued there'd have been a domino effect.

"I think he had such a strong sense of things that rather than be *right* and force the Fairchild people back, he decided that the development of solid-state physics would proceed better if there weren't impediments—if things could sprout elsewhere. I think he sublimated his own feelings to the long-term benefits of solid-state physics. I say that on reflection, and never having known him that well.

"People didn't love him as a personality. But they respected and admired him. In those days you could go to work for Shockley for a year as a technician, then go down the street and go to work for another company as an engineer, without a degree, and be paid as an engineer. All you had to do was work for a year for Shockley. That was the position he enjoyed in the industry.

"I'm sure I'm probably painting with too broad a brush.

Others will give different flavors to the story. But I believe I'm accurate from the point of view of spirit. I know I heard that conversation.''

And as for the issue that brought Morrow into Shockley's office, it was decided simply enough. Shockley asked Morrow his major. "I said it was physics. He turns to the guy in production and says I'm to stay in research."

Morrow completed his undergraduate degree in physics at Stanford and his master's in mathematics at the University of Oklahoma. "I came back to Berkeley to do work on a Ph.D. in mathematics, got tired of teaching calculus . . . because it got to be the same as making deviled-egg sandwiches. So I dropped out and tried to fashion some marketable skills, and ended up being in computers. About the time I had put in my apprenticeship in the computer area, the microprocessor revolution was upon us."

Indeed it was. In 1974, Intel introduced the 8080, the first widely used 8-bit microprocessor. It was expensive as components went, costing $360 each. But to be able to buy a computer on a single chip! The effect on people who understood the significance of that—as Morrow did—was immediate.

Within a matter of months, a company in Albuquerque, New Mexico, named MITS, a manufacturer of measurement and instrumentation kits, introduced an entire computer for sale in kit form, including power supply, memory and microprocessor, for $395. And they did it using a central processing unit that cost $360.

The MITS Altair 8800 had been nourished by a man named Les Solomon, the New York–based editor of *Popular Electronics* (now called *Computers and Electronics*). In early 1975, the Altair appeared on the cover of *Popular Electronics*. A revolution was underway.

"Can you imagine the effect on anybody technically oriented, or who ever wanted to own a computer? They saw at that

Personalizing the Computer

point they could have one. Why, you had a whole group of people that was already presold."

The notion of the computer coming out of its air-conditioned confines into the hands of individuals—a personal computer— was clearly one whose time had come. And the idea spread like blazes. On the West Coast, the first personal computer users' group, the Homebrew Computer Club, started in Palo Alto in the spring of 1975. On the East Coast that fall, in Peterborough, New Hampshire, a man named Wayne Green started a magazine called *Byte*, designed to be "the small systems journal." And everywhere in between, a crazy kind of excitement began to spread, from a hard-core coterie on out.

Here was a gizmo that, better than anything else, marked a synthesis between the buttoned-down mind of the design engineer and the freewheeling mentality of the free-spirited counterculture. A case could even be made that the emergence of the microcomputer, in the very months that saw the end of the Vietnam War, symbolized the reconciliation of two temperaments in the American mentality that had threatened civil rupture only six years before. The cool, impersonal computer that had represented established authoritarianism in the late '60s could now be in the hands of anyone with $395 to spend and a willingness to learn the steps necessary to communicate with the device.

Morrow was one of the first to pick up on the promise of the microcomputer and to help promote the new technology. Now, years after the fact, some credit him with exercising great foresight in being one of the pioneers of the microcomputer industry. "Nonsense! It was like living in the desert when they were laying tracks near my house. And one day the train came along and they said to me, 'Would you like to get on?' And I said, 'Sure, why not? Where the hell else have I got to go?' You're there at the right time. It hits you in the face. It's hard to say you shaped it. I felt there were opportunities there and I just poked around and looked.''

The view he saw from Berkeley, where he lived at the time, was different, Morrow feels, from the view he might have seen from Silicon Valley.

"We had more of a hippie, more of a university, more of an intellectual environment up here. But we were still close enough to benefit from all the creative aspects of the Valley . . . close enough to enjoy the benefits of instant access to the technical information there. But far enough away from the Mickey Mouse environment down there where you can't tell one company from another unless there's a sign in front, and where there is no sense of job identification or loyalty."

Morrow worked at the Center for Research and Management Science at Berkeley. With him there were Chuck Grant and Mark Greenberg. For a while the three of them talked about putting out a microcomputer kit of their own that would be marketed through a mail-order company in Oakland owned by a man named Bill Godbout. That kit never came to be, but that didn't stop the careers of any of the group. Grant and Greenberg would open a computer store called Kentucky Fried Computers, and later a microcomputer company called North Star Computers. Godbout's marketing company would grow and thrive. And another man who also worked at the research center, named Rodney Zaks, would later start the successful computer book publishing firm of Sybex. Morrow, too, would start a business of his own that would undergo a series of metamorphoses.

Besides these enterprises, other microcomputer companies were also starting up in the East Bay. IMSAI, in San Leandro, would soon become the biggest competitor of MITS. Among its key people, one would go on to start the ComputerLand chain of computer retail stores; another would start MicroPro, creators of WordStar, the popular word-processing package. Processor Technology of Emeryville would claim its Sol-20 (named after Les Solomon of *Popular Electronics*) was the

Personalizing the Computer

first *complete* small computer. In that same sphere was a young English writer named Adam Osborne whose *Introduction to Microcomputers* helped him establish his own publishing company, which he later sold to McGraw-Hill before starting his celebrated namesake computer company.

"The whole damned industry was over here then, which isn't to say there weren't some other companies in other parts of the country. Now let me tell you something about Silicon Valley then. They wouldn't even deign to sell parts to people like us then. 'For heaven's sake,' they'd say, 'don't tell anybody where you bought these parts.'

"We were a stepchild. They didn't care about these microcomputer companies, which they saw as just illegitimate children of the semiconductor industry. 'The hell with you people,' they said to us. 'We only need you now in this unfortunate '75 recession we're in.'

"I'll tell you something else. We co-opted them. We're now a legitimate part of it, but only because the market favored us. It isn't that we were magicians, but the market favored us, and none of *them* had the sense to see it coming. Their marketing skills down there in Silicon Valley? They're all adolescent! They focus about three inches in front of their noses. Anything beyond that, they don't see. And the reason the Japanese today are in the position they are, in semiconductors, is that they have a more mature attitude . . . they look beyond the next deal . . . they focus further into the future than the guys in the Valley. And if they're not careful down there in the Valley, they're going to lose everything in the process. If the Japanese dominate in semiconductors, they'll come to dominate in boards, and eventually in computers.

"If I had my way, I'd march these marketing guys in the Valley out in the parking lot every Friday night, thrash them all soundly, and march them back in, until they learned to look further than three inches past their noses."

How then does Morrow account for the phenomenal success of Apple Computer, the Silicon Valley company that, probably more than anything, brought Silicon Valley to the world's attention?

"From a design point of view, they [Steve Jobs and Steve Wozniak] had a great product. But the other important point is this, and this had to do with their having grown up in the Valley . . . they saw that you don't become a big company if you own all the stock. Over here, you see, we all own all the stock in our companies. At Apple, they sold pieces of the company in order to really move it. That is the atmosphere those guys came from, and it was more natural for them coming from there to fall into it than for the guys over here who were, for the most part, antiestablishment."

Unlike some who started in the microcomputer business in the mid-1970s, George Morrow is still very much in business, and while the story of his firm, Morrow Designs, isn't as well known as the now-legendary story of the creation of Apple in a garage by "the two Steves," the genesis of Morrow's company perhaps more truly reflects the overall growth of the microcomputer industry in the Bay Area and elsewhere: an evolving awareness among manufacturers of the users' interests, and a parallel growing awareness among the general public of what a personal computer is and isn't.

After an initial proposed venture that didn't get going, Morrow began marketing add-on memory boards for the Altair through a mail-order firm. True to form for Morrow, that soon took on the flavor of making deviled-egg sandwiches.

"I want a 'rush' out of life. I wanted to do some other products than the mail-order business was interested in doing. So one day, I said to my wife, 'What's the difference between my doing the products for someone else to market, and doing them for ourselves? Just an ad in a magazine, that's all.' It turns out there was a bit more than that, but in those days it wasn't a lot more. You put an ad in a magazine like *Byte* or *Popular Electronics*, and people sent you money in the mail. Some guys, like myself, didn't cash the checks until

Personalizing the Computer

we sent the product out. Some others took the money and maybe you never saw the product. The customers financed the products.''

Morrow did, indeed, start his own mail-order business, called Morrow's Micro-Stuff, to market his line of boards that could be added to small computer systems to enhance the capabilities of existing microcomputers. But he soon realized he would need a more marketable name if he was going to go after the consumer market. "We all thought in those days that this was a home-oriented market. We didn't realize then how important the business market was." So, in keeping with the tenor of the times, Morrow whimsically changed the name to Thinker Toys in 1976 and continued to market input/output, memory and S-100 bus boards by mail.

Some of the products were "real cash cows that produced a lot of money. There wasn't much competition, the overhead was low, product development was cheap, products were easy to define, there was no capital necessary to get into the business . . . it was all profit. You threw a product out, people gobbled it up. I didn't get rich instantly, however, because when you're not financed with venture capital, everything you make you push back into the business.''

By 1978, it was apparent to Morrow—and to the computer system manufacturers that were his customers—that however clever the company's name, it was fatal. The industry had started to become more business-oriented. Morrow's customers bought boards from him to make complete computer systems to sell to, say, a dentist who was looking to streamline his business.

"Now this dentist wants to look inside his new tool. He opens the top of the computer he's just bought from one of my customers. And here's a board in there that says Thinker Toys. 'Toys!' he says. 'I didn't buy a damn toy! What is this?' So my customers started to tell me, 'We love you, George, we love your products, it's all great, but the name is giving us some problems in the marketplace. Now you can either listen to that and flow with the marketplace, or you deserve whatever happens to you.' "

So in 1978, the name was changed again, this time to Morrow Designs. (Which is, not surprisingly, a pun in its own right, with "Designs" serving simultaneously as noun and verb.) At first, that company sold disk subsystems to computer builders, but in 1982, the company introduced the Micro Decision line of personal computers, whose low cost for performance is a wonderment of the entire industry. But because the company's annual sales are far less even than Apple's ad budget today, those sales are mostly through the word-of-mouth network that exists between knowledgeable computer users.

How does the iconoclastic, free-spirited George Morrow compare running a good-sized company doing $35 million a year in sales with the early days of Thinker Toys?

"A whole different skill set is necessary. Planning means more. It's a lot different from doing a product and throwing it out and hoping somebody will buy it, and if they don't . . . what the hell, throw another one out. That's different from today. And it's more expensive now. There's more risk, with more to lose. The industry is moving faster.

"The rules of selling a hundred or a thousand products a year, versus selling a million products a year, are totally different. You see, we first sold products to guys who were technically skilled enough to build and use them. But as you begin to get a universal product . . . who are we selling to now? To guys who don't know jazz. They only want to use it to do something else with it. It's a whole different set of customers. And if we want to keep selling, we have to keep finding new customers. And they're less sophisticated. They demand more for their money. So we either have to pull down the front-end barriers of technical expertise needed to operate the computer, or find new software for customers that makes it impossible for them to resist the hardware, the way VisiCalc did for office computers."

Personalizing the Computer

Yet even as Morrow came to learn more about the new market for micros, he found there was one big hole in his understanding. "I don't know what business I'm in. I haven't been able to figure it out."

Surely, the inscrutable Morrow is being droll, I figure. Using the Socratic method, he will, by a series of pointed questions, lead me to enlightenment about the real meaning of the microcomputer age.

Nothing doing. He means exactly what he says. And he proceeds to call my attention to how the telegraph companies failed to see that they were really in the communications business. And with that shortsightedness, he says, Western Union passed up the chance in 1886 to buy all the telephone patents for $40,000. The railroad companies, too, failed to see they were really in transportation, not railroading, and never had the insight to get involved in air transportation when that business was young.

"I read about all these myopic people, and I think, 'Dumb, dumb, they are so dumb. They didn't use their heads.' So I ask myself, what business am I in? The computer business? Nonsense! That's as crazy as saying you're in the telegraph business. Maybe we're in the productivity business, but that's too easy. I don't know. All I know is that I don't know for sure. Now I don't feel that those railroad guys were quite as dumb as I thought."

Whether he knows what business he's in or not, Morrow genuinely enjoys the work, and his emerging role as a colorful, articulate cult figure among computer savants. He has the satisfaction of knowing he is doing what he wants to be doing, earning the respect of his industry for doing it, and being rewarded for it, as well—though not rewarded to the extent he had until recently hoped. For just as his company was about to go public, the market interest in new and emerging microcomputer companies turned off. "I've got this big pile of chips that I can't get off the table. I didn't worry about getting them off the table until after I realized that I couldn't get them off the table. When you start out, you look for something to do that's going to produce some long-term growth. Now the operation's bigger and, what the hell, it looks like it's going to go on forever.'' And, for Morrow, for the time being that means staying right where he is. As for ten years out? That's anybody's guess, according to him.

The East Bay is no longer the hotbed of the microcomputer business. Rather, Silicon Valley has become the ''intellectual center, the hub'' of the microcomputer as well as the micro-electronics industries. Yet Morrow Designs remains in San Leandro, near the Oakland Airport.

"I'm close enough to Silicon Valley up here. It's like the difference between being a mathematician and an engineer who want to kiss a girl, but can only approach her by going half the distance each time. Ask the mathematician when he'll kiss her, he'll say, 'Never.' Ask the engineer and he'll say, 'Pretty soon for all practical purposes.' I'm close enough to Silicon Valley for all practical purposes.''

Up to this point, we have explored the technical, the financial and the historical aspects of Silicon Valley, whose nickname was coined in the early 1970s by one of the first local industry watchers, Don C. Hoefler. In the second half of the book, I want to look at the other dimensions of this place, to better understand it as a community in the fullest sense of the word: profiling the observers, lawyers, detectives, former convicts, artists, social critics and theologians who charge and are charged by this high-technology community.

5 | CIRCLE ONE: SCENE AT CLOSE HAND

14 | Valley Observers: Dick
Steinheimer, Artisan;
Elizabeth Horn, Chronicler
of Society; John C. Dvorak,
Journalist

The community of Silicon Valley can be
thought of as a series of concentric circles. At the core is
the central activity: the electron manipulating and deal backing that creates and sustains the place.

The next sphere out is made up of those who attend to this
core activity at close hand: accountants, marketeers, journalists,
lawyers, even private detectives.

Spinning farther out would be those who take the products
designed for technological applications and seek to pull from
them aesthetic creations: electronic artists and composers.

And farther still would be those who observe this scene and
are involved in it, but who, for reasons they will explain,
distance themselves from it: social critics.

If we want to understand where the technology of Silicon

Valley is leading the world, we need to explore these outer spheres of the society, as we have the core, to gain insights from a range of observer/participants.

There is also a distinction that needs to be emphasized. There is still a world that coexists here alongside the concentric spheres of Silicon Valley. And that is the Santa Clara Valley community. For many, perhaps most, of the residents of San Jose, Mountain View or Cupertino, Silicon Valley could as well be on the far side of the moon. Yes, there are streets here with names like Technology Drive. But there is also an intersection of Hope and Mercy streets in Mountain View. Certainly the people of the Santa Clara Valley are affected by the monstrous traffic jams and by the outlandish increases in property values brought about by the new technology companies that located here. But as men and women who run dry-cleaning establishments, bake bread, sell tickets in movie theaters or teach in primary schools, their lives remain relatively untouched by Silicon Valley.

At the point at which I am ready to move from the core to the first concentric outer sphere, to catch the experiences, observations and insights of people only slightly removed from the center of this place, I receive a rushed and breathless phone call from start-up mechanic Roy Dudley.

"Man, I just met the greatest product photographer! Fantastic stuff! I gotta use him. Can't afford him. Maybe, he tells me, we can work something out. When I saw his work, I thought of what you're doing, Tom. You gotta meet this guy. He's been around since the beginning. He took all the early product shots at Fairchild. Give him a call. Gotta run."

I have found, after several such conversations with him, that a phone call from Dudley requires a few seconds' breath-catching afterward. This one makes me pause even longer. Product photographer? Frankly, that job description doesn't sound as if it holds much promise as one that will help

Valley Observers

elucidate this place. What has someone seen who has spent all his time in a photo studio and a darkroom?

Anyway, I call Dick Steinheimer at his studio in Mountain View. In the course of explaining what I'm doing and why I'm calling, the comparison between Silicon Valley and Florence comes up.

"Well, gee," he says, "now that you've met the princes and heads of state, you should meet some of the artisans and craftsmen of the new Renaissance. And by the way, I think you'll find the equivalent of your medieval water-driven bellows in the Planar process. Both are basic technical achievements upon which huge developments rest. From the bellows it was coinage, which made possible the modern monetary system. Well, Planar was a basic manufacturing technique developed by Jean Hoerni at Fairchild, which became the only practical way to build reliable silicon transistors and integrated circuits. With the Planar process, it was as if you could take Coca-Cola and hold it in suspension and grow the glass bottle around the drink so it would be perfectly sealed and keep out all the impurities."

I arrange an appointment.

Stein, as he refers to himself, shares with another photographer a small office space and a large studio on a side street near the Bayshore Freeway. He is well over six feet tall, and has an engaging smile that puts a stranger immediately at ease. His aren't the features or habits of today's fashionable leather-jacket-and-jeans artisan. Rather, he looks as if he'd be more at home in a small town in the Southwest, sporting some silver and turquoise jewelry, selling his photos of desert sunrises.

The floor-to-ceiling bookshelf that takes up one wall of his office is dominated by books dealing not with technology but with railroading. It turns out that Stein has had four volumes of train photographs published.

The fascination with the iron horse is an avocation, however. Stein's profession is high-tech photographer. His professional work isn't on the bookshelf, because it doesn't appear in books. Instead, it's in slim marketing brochures and sales pieces he keeps filed away. Yet in spite of the overtly commercial context of the work, Stein and the other photographers of early microelectronic devices were creating a new photographic genre, a genre that would later become recognized by critics and the public as a legitimate art form when others created similar works to grace the covers and pages of *Scientific American*, *Omni* and *High Technology*.

Stein shows me a color photo he shot for the cover of a 1966 Fairchild employee-benefits brochure: the circuitry of a linear amplifier magnified 375 times. The design is incredibly simple by today's standards, but Stein's photographic image conveys the cool, compelling, geometric intensity of a Mondrian painting, probably never intended by the product's engineer.

Like many of the craftspeople who work here now, Stein backed into both the Silicon Valley and his pioneering role. But that's understandable. After all, the work done by electronics engineers is based on many years of formal study in that field. One doesn't "fall into" breakthrough work in engineering or physics. But job descriptions like high-tech photographer or writer simply didn't exist much before people like Stein. They not only created breakthrough work in their fields, they helped create the notion that there was work for them to do as craftsmen.

"I was working for the *Independent Journal* in San Rafael in 1962 as staff photographer when Fairchild Semiconductor opened a plant north of San Rafael to make diodes. I didn't know what a diode was, I didn't know what a semiconductor was, and I didn't know what a Fairchild was. But I got an interesting assignment one day to go up and photograph the plant for a news feature."

Fairchild was so impressed with the resulting photos that Stein was offered a job at the company's Mountain View headquarters. It was the beginning of a new world for him when he took the job. Not only were the subject matter and the environment new but, coming out of newspapers where resources were tight, and "fed up with the stodginess," he was pleasantly astonished to be given the then huge sum of $5,000 by Fairchild to go out and buy whatever equipment and supplies he needed to start and build a photo department.

If the subject matter and the budgets were new to the former photojournalist, so was the pressure. And the exhilaration.

"My boss, John Hall, used to spend maybe an hour or two a day with Tom Bay, the director of marketing. By his sitting around with Tom, telling jokes, talking things over, we got jobs into our marketing services organization that came right from the top, just one voice removed from Bob Noyce, the general manager. So that we were not at the tail end of any idiotic chain of command.

"Many times on Friday afternoons, at five or six, we'd get the biggest job of the week. That meant working the weekend on a series of photographs, or a brochure. Or it could be an urgent need for support material for a salesman making an important call in New York on Monday.

"The creative project planning done by our impromptu interdisciplinary teams, with a minimum number of people, was great! Since then, I've found that people only fall into that format when their asses are hanging from a wire.

"The job was such a release for me. Here we had a field with all kinds of subjects, of people, of processes, of products . . . with very few pictures ever taken of them. For a person like myself, who liked to experiment, liked to take things and try to deduce some of their meanings by photographing them . . . wow! It was fantastic!"

One of the perks of being a craftsman, in Silicon Valley or anywhere, is the chance to sit in unobtrusively and watch as the "heads of state" conduct the affairs of state. Stein was encouraged to walk around the company with a camera, documenting life in the growing enterprise. He had a chance to observe from the inside the development not just of that company, but of a whole new industry.

"Bob Noyce was a delight. Approachable, not always talkative. His and Jack Kilby's patents for the integrated circuit make them, in my mind, future Nobel laureates. Once, there was a very serious meeting with customers, and during the meeting Bob's secretary came in and handed him a little written message. Just real smoothly, he got up and said, 'Well, gentlemen, I think I'm going to have to go now, but I'm sure that you can conclude the meeting.'

"One of the salesmen looked out the window to see him drive off as if he was going on a mission to the President. The salesman was awfully curious, because he had seen Bob throw the wadded note in the trash basket. So he went over and opened it and it said, 'Your wife just called and wants you to come home immediately. The burro has gotten out again.' "

"I had my lab over at the first real Fairchild plant, which was on Charleston, by San Antonio. The story has it that the power company or the city hadn't hooked up the electrical system on the day the plant was supposed to open, yet the people were there ready to work. So Bob Noyce and Gordon Moore went out and hooked up the high-voltage wires to the building. I don't doubt the story. It's funny to think of the great Ph.D.s out there doing that."

There were so many unassuming men of extraordinary accomplishment then, according to Stein. Like Steve Ammann, a quiet engineer who went modestly about his work in a back lab in the instrumentation division at Fairchild. "All the guy ever did was lay the groundwork for today's digital voltmeter. Well, wow! That's like inventing bread. These inventions were coming up around us like that all the time."

*   *   *

Valley Observers

Chief among Fairchild products was the integrated circuit. Whereas the transistor is essentially a simple on-off switch, a number of which reside on a printed circuit board, the integrated circuit put all that switching ability on a single fleck of silicon. And Fairchild not only had the technical ingenuity to develop and mass-produce such products, but also had the savvy to market them to potential customers using an in-house group that built and demonstrated sample applications designed around the products.

"We had a man there and one day he came in and wanted me to do a picture. In his hand he had this little box. It was the world's first totally solid-state radar. We did application notes on it and gave that technology to the world. We said, 'Look, you can use Fairchild integrated circuits and build a radar device that doesn't require a ten-ton truck. You can hold it in one hand. The first transistorized televisions were designed at Fairchild in the applications lab, and the designs were given to the television companies."

The military was the biggest customer for electronic components in the early days. For them, the high prices of the first devices were no obstacle. In the commercial/consumer world, however, such high prices would ruin the competitive edge of a manufacturer who chose to incorporate the new technology, so its acceptance in nonmilitary applications was slow. Yet, in a Catch-22 situation, until there was a mass market for these devices, their costs would never come down through mass production.

"Tom Bay's the guy who directed Fairchild to bomb the price of ICs and launched the commercial and consumer business for integrated electronics. Bay's price-cutting ads appeared in *Electronic News* one day, and with that the capabilities of electronic devices were made affordable to manufacturers of such consumer products as home appliances, sound systems and automobiles. The world changed after that."

* * *

"In the mid-'60s at R&D, Frank Wanlass and Phil Ferguson were among the key people who developed MOS [metal-oxide-silicon] technology, a variation of the Planar process. Then one day they left, to start General Micro-Electronics.

"Do you realize in retrospect what that meant? It was the beginning of the technical revolution that led directly to today's extremely powerful VLSI [very-large-scale integrated circuit] chips. It's as if some Neanderthals were sitting around a cave and invented fire, and a couple of them got up and left to start Cro-Magnon civilization."

Stein not only saw modest men perform signal accomplishments. He saw the effects on their families, too.

"There were widows, Fairchild widows, who supported each other. The work often became a sort of mistress for the husbands. Wives would get together socially because they wouldn't see their husbands for periods of time, ranging from nights to weeks, often because they would be traveling. And even though men would leave the sales force and go into management, they would still continue to operate in the same kind of nonstop format.

"This revolution wasn't based on illicit sex. In a sense, the challenges and excitement of the field replaced sex. The real mistress was this incredible opportunity in a brand-new field, working with bright people and tremendous challenges. When you have these inventions coming through your hands and through your head that are world changers, day after day, literally opportunities of a lifetime . . . it can be distracting."

"The achievements of some of these men in the Valley . . . there was probably nothing that ever prepared them for being as successful as they became. It's as if they had been touched and converted into gods. When you talk to them, the qualities you like are still there, but because of their enormous rise to positions of responsibility, they can't help but be controlled

Valley Observers

by it, too. And so it's difficult sometimes dealing with the gods, though the friendships that they have accumulated on the way up are typically very real. Beyond those friendships, however, they have responsibilities to friends, family, stockholders, employees, the future; that is an unimaginable burden.

"Not only that, there are still active gods who cannot let misinterpretations come between them and their stockholders, friends or employees. They must be careful and find the 'right' answer for each question. But for each right answer, there are a million other answers, many more interesting, and some more correct."

In 1970, Stein and art director Lawrence Bender left Fairchild and went into business on their own. Why should the craftsmen be any different from the princes? It was while looking for a commercial photographer that Roy Dudley recently crossed his path.

Among Stein's clients over the years is Intel. "All I have to do to come across as old as Father Time is to say I worked with Intel people in 1971 on the first brochure and film on microprocessors. There's a funny, amazed reaction. You see, most engineers today have never known a world without microprocessors."

• • •

Dick Steinheimer represents the experiences and observations of a craftsman who worked in a hardware-manufacturing enterprise in the early days. Elizabeth Horn, on the other hand, has seen life among software programmers today.

She was, when we met, marketing director of a mid-sized company in the Valley that creates and markets software packages for large companies that still use the huge, room-filling mainframe computers. It may come as a surprise to people used to paying perhaps a few hundred dollars for their microcomputing software to know that a single package for the rarefied ether of that mainframe universe can start at $50,000.

Among her responsibilities, Horn oversaw a large staff of marketing specialists, as well as a complete printing shop, whose collective goal was to promote sales of the software written elsewhere in the building. Horn also contracted out a number of jobs to advertising and public relations agencies, and to freelance writers, designers and video artists. It's a responsible position for one still in her twenties, although like many others here she didn't set out after the position when she graduated from college in the Midwest.

Like many other English majors, she intended to become a teacher, but, like many other English majors, she soon found she was a glut on the market. So, with some background as an artist, she secured a job in the graphics department of a computer software company in Cincinnati doing layouts and paste-ups. That was her entree to the business. Before long, she was writing speeches for the company president.

She fell in love with California from a distance and decided to ''make a big jump, out of a marriage and out of a job, and come here. My first move away from home, trying to figure out what I was going to do with my life.''

Eventually, she got a job writing promotional copy at this firm, and within two years she had been promoted from writer to writers' manager.

Horn is a tall, statuesque woman, with a mane of reddish hair and a preference for stylish tailored suits. She turns heads when she enters a room. Not atypically in this county where, statistically, there are more split-ups than marriages, she has been divorced twice.

The place where we agreed to meet is the Decathlon Club, which, as much as anything, is *the* club in the Valley. A company knows it has passed the raw fledgling stage and reached a certain level of stability when it can take out a corporate membership there. The Decathlon exudes what might be called the understated elegance of Silicon Valley. A stream

Valley Observers

enters the clubhouse in the rear, spills over a waterfall—a small one—and meanders through the entire club, past the multilevel bar and restaurant and under a bridge or two, before it passes out the front door. But, hey, it's an understated, tasteful stream. The water isn't artificially colored or anything.

Horn has worked among hardware engineers as well as software programmers. I wondered what differences she'd see between the two. "Software people sit by themselves and do head things at the terminals. Hardware people are constantly collaborating with other people, so they seem more outgoing, a little bit more balanced."

Horn finds the programmers fall into two categories. "There are those who work pretty hard and are pretty standard. They come in about nine and leave at six. Then you've got the incredible programmers, the ones who do the work of ten people. Oftentimes, they bring bedrolls, they sleep at work, they eat at work. A lot of them never leave their offices for days. Then you might not see them for several days."

Those are the programmers that interest Horn, both as professional colleagues and as people. "They have lots of fast cars, unfortunately a lot of drugs, and a lot of women. Many of them tend to be great personalities as well. They are witty, well-rounded, have lots and lots of knowledge that hasn't come from college. They have a charm that . . . at times it's frightening because they are so quick, and the drugs can make them so incredibly quick, that many people can't tolerate being around them at all."

One such programmer friend jokes that he is going to start a club for white males in Silicon Valley called the Unoppressed Minority. His slogan is "We don't have anything to bitch about and we're angry as hell!"

"He's always doing monologues in bars about the difficulty of being a white, single, successful male in Silicon Valley and getting absolutely anything he wants. Where is there to go from there? But I see him at times, too, when he gets so frustrated at the slowness of everyone else around him. So he sits back down at his terminal and he plays with his own brain on the screen."

There are extraverted programmers, and there are introverts. "We had one programmer in our office. He was considered one of those brilliant types who was capable of doing the work of ten men, writing sophisticated programs to enlarge the capabilities of big hardware systems. He was always very well dressed and extremely quiet. Another guy shared his office and we would go and have parties back there. We'd turn out the lights and pull out the guitars, light the Sterno and put it on top of the CRTs, and we'd start drinking French champagne. We'd do this for hours and this other man would never look up from his work at his terminal . . . even with fifty noisy, partying people jammed in his office.

"His indifference became a joke. How are we going to shake him off his position? we wondered. Well, one day I went into his office and I was startled to discover that he was not typing away at the terminal. He had smoked about twenty cigarettes down to the butt end, and stood them up like sentries around his keyboard. They were all smoldering like little birthday candles. And he was just sitting there staring at them. A few days later he was gone. He'd left." The isolation of a man-terminal relationship, to the exclusion of the surrounding communal world, finally got to be too much. "A lot of people like that just kind of dissipate into a fine mist."

However *au courant* Silicon Valley might appear with respect to equality of various kinds, it is still very much a male-dominated world. In terms of sheer numbers, that may not be the case—women are more numerous in production-line work—but it certainly is true in professional jobs. The few professional areas in which women seem to have made the greatest inroads are marketing and research, far ahead of such fields as finance, operations, manufacturing, engineering or *ceo*-ing.

"I was terribly limited in my company in Cincinnati. I

could not be a manager because of the distrust there among the men in the company for women, though on an informal level I was a manager.'' As noted, Horn became a manager with over twenty people reporting to her within two years of starting this job.

''There are women who are very angry about what they call lack of opportunity in this industry, the low number of women who are software programmers, for example, or physicists or working on gene splicing or hardware. I believe if more women get interested in those professions—any professions, really—the opportunity exists here. There just have to be women who are willing to step into the opportunity. Some people disagree with me on that.''

A good part of the social life in Silicon Valley takes place in bars. And one can discover as much about life in this community there as on the job.

''There's a lot of drinking in the Valley. Some people go into a bar and stay there eight or ten hours, night after night. Incredible stamina is a prerequisite to this sort of activity, but it is a good way to learn about who is doing what to whom, and who has moved where from which company. The Valley bar scene promotes a lot of incestuous interpersonal relationships, too, because you get the same crowd all hanging out together and eventually it's like a big communal family where everyone has sort of slept with everyone else. It gets a little bit claustrophobic.

''A lot of editors go into bars seeking information . . . it makes for a good deal of intrigue. Some salespeople I know have been trashed for some of the things they have accidentally said in public about when a machine is coming out, or why a machine that's supposed to be coming out isn't coming out. Rumors do fly in bars. And mostly from men, interestingly enough. I guess it's because men are still the predominant movers and shakers and they do stand around and talk about things. Most of the women in the bars are looking for these movers and shakers or the venture capitalists. They are shopping, which is disappointing to me, because there aren't that many women that you can really talk to in these situations. "I spent most of my time in the bar scene conversing with the clever, entertaining men. I wanted to understand how they lived. They always seemed to have numbers of women hanging around them. These gentlemen often boast about their 'stables of women.' One of them claimed that once he'd branded someone she could never escape him. It's kind of funny that most of the women who are 'groupies' of a particular individual eventually all get to know each other and get along quite well together . . . almost like a harem. Which is really curious when you think about it. Kind of like users' groups."

One of the things that lured Horn to California several years ago was the fact that it is the headquarters of the business that is really closest to her heart: the motion picture industry. Her interest in films has already led to her involvement in a Public Broadcasting series on Silicon Valley, and that further encouraged her to write a documentary filmscript on the subject for which she is now seeking financing. She feels that her involvement in the corporate world has given her the business sense she needs to sell and produce her ideas for films.

Another project very close to her heart is a filmed biography of a "tempestuous, crazy, revolutionary, wonderful character" named Maud Gonne—the muse of William Butler Yeats and a celebrated Irish patriot during the last years of English domination.

"The thing that interests me most about her was the great romance with Yeats that was never consummated, though for thirty years it was really heated. The more I found out about her, the more I loved her. She was a character who traveled with monkeys and birds. She was concerned with teaching women to clean their firearms correctly so they could be ready to get involved in the Irish cause. She took a bouquet

Valley Observers

of flowers with a revolver in it to a woman, a revolutionary, who was in prison. She was thrown in jail herself for that. Her life was very colorful and romantic.''

And what on earth would be the appeal of a rebellious Irish patriot to a Silicon Valley marketeer? "Being in the midst of all this technology, romance is very important to me. There's so little of it here . . . and there are so many machines.''

However much Silicon Valley may originally have been only a stopping-off point for Elizabeth on her way from Cincinnati to Hollywood, and however much she still wants to pursue a career in films, she's found this place is beginning to cast a spell on her in its own right.

"There's a seductiveness about Silicon Valley. I'm intensely curious about what's happening to Apple, to Gene Amdahl and Trilogy, and what's going to happen to Altos, and to Osborne. The natural drama that exists here is compelling to me. I find I don't want to give that up. I want to write about what's going on here and about these people. I think it's important . . . historical . . . fantastic . . . scientific. It's the stuff that epics are made of!

"One of the things that first impressed me when I arrived here was how the people in Silicon Valley can be classified as brilliant, very brilliant or just extraordinarily brilliant.

"You get the brightest of the brightest psychologists and sociologists and musicians and video artists and designers that are trying to get all this stuff going. Everyone is sort of careening into this Valley, all at the same time, and it's all happening so fast and deadlines are so short and things are changing so constantly . . . it's like this hum of activity. When some of my friends from L.A. come up here to visit they say, 'How can you stand it? It feels as if it's going to implode!' And it's true. You can feel it. You can feel the tension and the pressure, and I happen to thrive on that.''

Yet she sees shadows over the Valley, as well. "It's alarming that so many people here don't want to wander out of their territory. They don't care what's going on in the Midwest. They don't want to go to Paris. There's so much happening here that they get tunnel vision *extraordinaire*. You get presidents of companies who start narrowing down their focus until it gets so that if you talk about anything else for any length of time, they get real antsy, as if you're wasting their time. That's scary.''

Shortly after our meeting, Horn left her job to open her own film production company, trying to raise money to make her Silicon Valley film, supporting herself by doing marketing consulting work. ''If you're not busy building your own dream, you're busy building someone else's.''

• • •

John C. Dvorak (''Try to use the C. Every time the C has been left out, some weird little mishap occurs'') is the Consulting Editor (''It's a title I created myself. Think about it. What does it mean?'') of *InfoWorld*, a newsweekly for microcomputer users. He is also the author of ''Inside Track,'' the publication's gossip column studied avidly by its 100,000-plus readers.

Although the newsmagazine is headquartered in Menlo Park, Dvorak does his editorial work, and writes his books, in a rented office just north of Berkeley. ''They have the best gourmet stores here, and I'm probably the best cook in Silicon Valley.''

Though the column might occasionally refute it, Dvorak's conversation tends more to the droll than the outrageous. As an editor of a weekly publication, he has to deliver fast-turnaround copy that covers a passing scene. For example, from the June 4, 1984, *InfoWorld*:

Last on the list was the parody of ''Inside Track'' found buried in the back of the April issue of *Creative Computing*. Unfortunately, it portrayed me as someone interested in **expensive** clothes, **Mandarin** duck, **Bur-**

gundy wines, **buxom** blondes, and **fast** cars, rather than in microcomputers. Actually, I'm interested in expensive clothes, Mandarin duck, Burgundy wines, buxom blondes, fast cars, AND microcomputers! Who isn't?

Prediction Dept., Irony Subdept. Within the next few weeks, you'll be surprised when VisiCorp makes an expected announcement before anyone expected it.

More Obnoxious Language Dept. I was talking to the women who heads the *New York Times* syndicate. She's going to syndicate a column that introduces people to microcomputers and, according to her, is "**reader-friendly.**"

It will highlight important uses for microcomputers, like recipe filing. **Let's all join hands and try to contact the living!**

His own reading tastes, however, run to authors of works that are serious, considered pieces on the role of technology in the world: Marshall McLuhan, Lewis Mumford, Arnold Toynbee and the French philosopher Jacques Ellul. At the time we met, he was finishing up his own book on the sociological implications of machine intelligence.

In the course of a conversation, Dvorak seems to move, chameleonlike, between several roles: the scholarly, detached observer; the college student out on a lark to prick the goad of a major industry; the defender of an establishment-coming-into-being. Yet he seems at ease in each role.

A native of the East Bay community of Newark, and a graduate of Berkeley with a degree in history and chemistry, Dvorak happened into the world of microelectronics quite by accident. He was in the basement of the Lawrence Hall of Science in Berkeley one day in 1975 when he saw "all these kids with these computers. It dawned on me, I'm going to be working for these punks if I don't find out what they're doing." Soon thereafter, he bought a microcomputer, started a newsletter called *The Software Review*, and created a company called California Software to sell public-domain packages.

Because of this experience and because he had done some occasional writing for *InfoWorld*, he was offered the job as editor of the publication when an opening occurred.

Four months into the job, "I figured since I was editor, I would just drop a gossip column in. I knew I had the contacts to do one. So one day, 'Inside Track' shows up. It was a different style of writing than I normally do.

"I developed this outrageous technique using boldface adjectives. The thing is written more like advertising copy than journalism. It's really heavily read. And I think it has a lot to do with the fact that readers don't know that they are reading ad copy. I think one of the reasons it's so popular in the industry is because most of them really aren't literary at all. But when they read ad copy it makes them feel real comfortable."

Typical material for the column has to do with the comings and goings of Valley and industry celebrities, as they travel from one company to another (which certainly gives Dvorak job security), and speculation on what companies plan to do or wish they had done. Regarding that old gossip-column standby, affairs of the heart, Dvorak has a hands-off attitude.

"It's kind of a taboo subject because the industry is so filled with entrepreneurs, and entrepreneurial activity always causes terrible interpersonal relationships. I think it would just be a little too hurtful to even bring it in."

Then why is the column so popular? If it doesn't report on purely technical bits and bytes on one hand, or fits and fights on the other, what news is left?

"Silicon Valley is a close-knit community. It's no different from a farming community. Everybody shifts jobs so much that it's almost like a high school scene where you know everybody because they were in one of your classes at some time or another. And there's this voracious appetite for information because there might be a trend that's happening that could ruin your business.

Valley Observers

"The other thing is the gossip column is actually entertainment, and there is very little entertainment in Silicon Valley, writing-wise. The articles are dry. They tend to be trade. There's very little humor, and people in the technical industry like humor. They don't like to admit it, but they do."

To Dvorak, Silicon Valley is "a milieu that has its own value system. It's really based on creativity in a very peculiar way, a funny kind of creativity. It's starting little companies up and selling them and going off and doing another one."

And given the value system of this milieu, it's sometimes difficult for a journalist to cover the place, since "journalists tend to have left-wing tendencies.

"The milieu has a certain way of thinking, and that's the way it is. You go in there and if you're bucking that thought pattern, because of your own personal prejudices, you're going to get nowhere. If you have preconceived notions about wealth or why these guys are rich or the fact that some of them are employing illegal aliens, for example, you'll never find out anything from them. They are going to sense that. Most of these guys are extremely sensitive. I mean, that's one of the parts of the game."

Ironically, considering he is the gossip columnist of the place, Dvorak doesn't really think Silicon Valley is all that exciting. "It's tame. It's ridiculous. It's almost boring. If anything, these guys become a little flamboyant, but it takes a very standard approach. You get a Ferrari, a red one. It's always red. And drive it around and grow a big mustache. Big deal! You know, it's not as though anybody is walking around in robes or anything. There's more weird people here in Berkeley than there ever will be in Silicon Valley.

"We have these big trade shows down in Vegas, and all the guys are there. And with the exception of one or two men who are gamblers, nobody gambles. They just hang around the bar at night and they all bullshit about products. I think if you want to epitomize Silicon Valley, just walk through downtown Palo Alto any day of the week. Beautiful place. The weather is fantastic. But it is death! The place is so *boring*!''

Stein, Horn and Dvorak offer very different perspectives on the same place: Stein's exhilarated sense of having observed ''gods'' at close hand, Horn's sense that it merits film treatment on a grand scale, and Dvorak's professed uncovering of only boredom.

Do they differ in outlook? Clearly. Do they contradict each other? Not necessarily. They only examine various facets of the same environment: Paradox Valley.

No one voice, no omniscient narrator, no single guide could pretend to be the Baedeker, the Virgil here. But all the voices together will, hopefully, convey a sense of this place and lead, perhaps, to some comprehension of what it means and where it's taking us all.

15 | The Law—Observance and Breach: James Pooley, Attorney; Bob McDiarmid, Private Detective; John Draper, 'Cap'n Crunch'

One way to look at the central activity—the obsession—of this place is to see it as breaking out, whether it is breaking through older limitations in a technical field, or breaking free from a current position as an employee into a new position as owner. That drive goes back to 1954, when William Shockley left Bell Labs to pursue the commercial opportunities offered by his transistor. Every time a protégé steps out, a mentor is left behind. Shockley is said to have referred to those who left him to start Fairchild Semiconductor as "the traitorous eight." His company never got over the loss.

And Shockley begat Fairchild, which company in turn was hurt when Noyce and Moore went off to form Intel. And Intel would later bring suit against a spin-off called Seeq for damages. Explore nearly every case of corporate begetting and you'll find feelings of individual betrayal and assertion.

The former boss contends, "I taught him all he knows," or "She pirated my best employees and my customer lists when she went off on her own," or "The bastard stole all my trade secrets!"

The departing employee responds, "You can't keep me here against my will, and you can't prevent me from making a living. I've got a right to move on and move up. That's the American way."

Is it any surprise then that the new technologies have affected the law, testing the flexibility of that ancient codification of fairness, stretching its ability to arbitrate in unprecedented situations? It's one thing to hand down a verdict when there's a smoking pistol. It's more difficult for a layman—judge or juror—to do that, when an entrepreneur is charged by his former employer with theft of trade secrets, and the case revolves around a fine point in processing technology that is fully understood by perhaps only six people in the world. The challenge of doing justice is even more formidable in the case of criminal technology theft, where guilt must be proved "beyond a reasonable doubt."

A new and very active area of law has opened up to formalize the gavotte that entrepreneurs and employers perform with each other. But however complex the litigation, it is, at its root, a formal expression of a personal clash between a sense of betrayal and a declaration of independence.

James Pooley is one of the clean-cut, smooth-featured young men who typify the new generation of movers, shakers and counselors to the same, in Silicon Valley: successful in a traditional profession, but not cowed by authority or tradition. He grew up in Wilmington, Delaware, and, like many of his contemporaries, got his first exposure to the practice of law by watching Perry Mason on television as a child. "When you're at that age, and you're not really sure of what you

The Law—Observance and Breach

want to do, but you think you want to do something important, law seems like a logical thing.''

Developing an early interest in international law, he pursued his studies at Columbia. But he soon found that an appointment to the Paris office of a Wall Street firm took somewhere between ten years and forever. Worse yet, international law put him to sleep. During a trip to California, he fell in love with the place and determined that he would like to settle here. He managed to get a summer clerkship in 1972 with a small Palo Alto law firm, expecting that litigation on the Peninsula might have to do with ranch and farm law. Instead, he found the chief activity was counseling new companies. Entrepreneurs would come in and say, ''We need to become incorporated, and . . . oh, by the way, our former employer is coming after us, claiming we stole trade secrets.'' Or customer lists, or employees.

Pooley was offered a position there upon graduation, and as a result of the firm's concentration on start-up business, he was exposed very quickly to considerable trade-secret litigation.

Why should lawyers be different from their clients? A disagreement between two founding partners in that firm prompted one to spin off. Pooley and several other attorneys left with him in 1978.

The offices of that spin-off firm—Mosher, Pooley, Sullivan and Hendren—are located on the fourteenth floor of an office building in downtown Palo Alto, overlooking the Valley of the Entrepreneur's Delight. Actually, it might better be described as the Valley of the Lawyer's Delight, since, according to Pooley, ''there are more lawyers per capita in Palo Alto than in any other municipality in the world, primarily as a function of the technology industry.''

In spite of the fact that Pooley had no technical background—''I was a social science bullshitter all the way through college''—he has, in the course of his work, become an authority on technology law. He is the author of a book called *Trade Secrets: How to Protect your Ideas and Assets* (Berkeley:

Osborne/McGraw-Hill, 1982), and past chairman of the American Electronics Association's Lawyers Committee.

As far as his lack of formal technical training, he finds it a benefit because "the people you are selling to, judges and juries, are similarly not technologically literate."

About 70 percent of his practice has to do with defending new companies that are getting started. Yet since some of these successful start-ups spawn spin-offs themselves, Pooley finds himself on the plaintiff's side about 30 percent of the time.

The fundamental question in all this litigation is who owns what in the intellectual property area. How much engineering insight is native ability, and how much was taken from the job?

"I haven't seen a company yet in the Valley that doesn't encourage people to take work home. So it's not unusual for engineers to have stacks of stuff lying around. A lot of these people tend to be a little bit disorganized because they are such geniuses, and they think, 'Oh, that material's not worth anything, that's old.' And they will leave to start a company, or even take another job, and become the subject of a lawsuit, and all of a sudden the court tells them to produce this stuff. It can really look bad."

The incidence of this kind of litigation is increasing, and for several reasons, according to Pooley.

Chief among them is the fact that more intense competition, compared with fifteen years ago, means that success in the electronics industry is "predicated on developing a technology edge which allows a company to hit a market opportunity that is moving faster and getting smaller." In a climate like this, companies realize they must protect that "edge" as much as they protect their physical facilities and inventory.

"Furthermore, there are not too many totally new technologies. We're slicing hairs thinner and thinner. It's difficult for an employee to leave a highly specialized job and go

The Law—Observance and Breach

apply his skills and talents elsewhere without coming close to violating the employer's trade secrets.''

Another reason for increased litigation is that ''some companies perceive the lawsuit as a means to eliminate competition, by scaring away sources of financing, potential customers, suppliers, and so on, essentially crippling a new business before it gets off the ground.''

While 90 percent of the cases never go to trial, the process of litigation wreaks havoc with the new company, diverting its efforts when it can least afford the distraction.

The upshot of it all is that even if the defendant entrepreneur is squeaky clean, he may be forced to settle and pay just to get the damn thing over with.

There's always the risk, too, that even if the case goes to court, the judge will be out of his depth. There was one situation in which a Superior Court judge in Palo Alto issuing a $1.5 million judgment against a defendant said, in effect: ''I don't know what the trade secret is but it's around here somewhere, and I think you've used it.''

Perhaps the most famous theft-of-trade-secrets case in Silicon Valley occurred in Santa Clara in June 1982, when two employees of the giant Japanese computer firm Hitachi, Ltd., were caught buying proprietary IBM information by FBI agents in what has been described as a ''sting'' operation.

At first Hitachi sought to have the charges dropped, saying that it was IBM, not the FBI, that was behind the bust, in the interests of stopping a competitor rather than enforcing the law. The Japanese firm later changed its plea to guilty.[12] Hitachi had paid $600,000 for stolen design workbooks and assorted hardware and software. The company and several employees were fined over $25,000 and the company faced payment to IBM of $300 million as reported in the *Wall Street Journal* in November 1983. (The amount of payment has since been reduced.)[13] Hitachi is also reported to have agreed to allow IBM to review its new products for the next five-year period.

However little IBM may have been involved in the sting, the case certainly sent a message to all competing computer companies who might think of trafficking in IBM Confidential material.

Yet while in that case IBM was the aggrieved party, there are, according to Pooley (who represented another company named in the case), instances where the Blue Giant uses the courts not against corporate equals like Hitachi but against individual employees-turned-entrepreneurs, sending a message to all other employees.

In one case, IBM brought suit against two men, whom Pooley represented, in a trade-secret matter which was settled out of court. "The defendants agreed to the settlement only because they realized that IBM could make them both unemployable and bankrupt simply by keeping it going. Their only 'sin' was to *plan* to start a legitimate business of their own. But when IBM summarily fired them and cranked up the war machine, they had no choice but to settle.

"When we asked IBM, after the lawsuit was filed, to tell us what 'trade secrets' it was concerned about, so that the two scientists could agree to steer clear of that area, it refused, forcing us through weeks of horrendously expensive litigation. Why? I can only conclude that IBM's primary objective was not to protect its technology, but to bring these two individuals to their knees, so that IBM could dictate the terms of a settlement that could be waved as a warning flag to its other research scientists."

However much IBM might disagree with that assessment, the fact is that the company's legal resources are formidable. In fact, Frank Cary, chairman of IBM's executive committee, was quoted in *Electronic Business* once as saying, "In our organization, the legal department is the only department with an unlimited budget . . . which it exceeds."[14]

IBM's pursuit doesn't stop simply with former employees. Subsequent to our interview, IBM brought suit against Jim Pooley and his fourteen-person firm, claiming they had caused

The Law—Observance and Breach

IBM, which does about $40 billion a year in sales, "irreparable harm."

The suit, filed in Santa Clara Superior Court in late 1983, says Pooley divulged proprietary IBM information to potential backers of the two men mentioned above.

Someone close to one of those who supposedly saw the alleged trade-secret information claims it was nothing that was not already common industry knowledge.[15] Pooley claims this is only further evidence of IBM's determination to intimidate prospective or actual entrepreneurs, now by extending the threat of legal action to include even the entrepreneurs' attorneys, and possibly other providers of services. (In April 1984, IBM's application for an injunction was denied. The case goes on.)

What can the employee-with-the-soul-of-an-entrepreneur, no matter whom he works for, do to minimize the risk of legal action? "You can maintain a record of the fact that you have not crossed over the line in your plans, to define what belongs to the employer and what doesn't, seek counsel on what general skills and knowledge you can look forward to taking with you, and not solicit employees or customers before you leave."

The effort to remain, and to give the impression to the employer and the courts of remaining, aboveboard is important because so many of the lawsuits, especially in this environment, are generated by emotion, by raw gut reaction.

"It's like a divorce. There are feelings that this is an act of disloyalty. The employer taught the new entrepreneur everything he knows, right? And then the guy goes to use it against him. I know of one case in which an employer tried to exploit a personal relationship one of his female employees had with someone who left. He made her go around carrying a tape recorder in her purse."

Emotion exists on the defendant's side, too. "Typically, they react with disbelief that anyone can, in essence, accuse them of stealing. For many of these people it's their first experience with litigation. They get a complaint which basically says they are thieves, and asks for more money than they ever dreamed they would have access to. You know, often they are just numbed by this event, and they are scared as hell.''

Companies can also minimize their exposure to such risks of infodrain: by getting signed agreements whereby the employee acknowledges he understands something is proprietary information; by educating employees and explaining who owns what and why; by distributing reminder memos and posters around the company; by tight control over document possession and copying, including use of colored candystriped borders that appear as black on photocopies; and by periodic checks on outgoing briefcases.

All this would seem like common sense, but it isn't always done. In one celebrated case which Pooley recounts, Motorola brought suit against some employees who had left to join Fairchild, claiming there were trade secrets at stake. The judge denied relief to Motorola, saying it had never told the defendants what were considered trade secrets and, second, had frequently let outside visitors in to see the process in question, without asking them to sign nondisclosure agreements. As Pooley recalls it, the judge, in effect, said to Motorola: ''You're not just asking me to close the barn door after the horse is gone, but you're asking me to build the barn for you. And I'm not going to do that.''

Still, all the precautions in the world can't keep trade-secret information perfectly secure. Only the most draconian measures, unfailingly enforced, will ever keep technical information from going for a walk. And that clearly is not in the temperament, or the best interests, of the industry. Indeed, putting a tight clamp on the flow of information would probably, in short order, cripple the entire information-processing industry.

The Law—Observance and Breach

Trade secrets can be gained in other ways too, all open and aboveboard. The average Silicon Valley bar after five in the evening contains more information on leading-edge electronics than most university libraries.

"If you go down to any of the bars in Cupertino on a Friday night, seat yourself at a table near a bunch of guys with pocket protectors and just lean over, you can probably hear a great deal of proprietary information."

A common type of usually legal intercompany industrial espionage occurs when Company A offers a higher salary or shares of stock to lure an unhappy employee out of the research group at Company B. (Although if Company B feels sufficiently injured, it will charge Company A with having made the hire only to get access to its trade secrets.)

Information can pass internationally as easily as between companies. "Unfortunately, that is easy pickings here in the Valley," says Pooley. The Japanese, in particular, have earned a certain reputation for exploiting to the utmost a characteristic carelessness here in Sieve Valley. Their "listening posts" have been here a long time, and there is plenty of technically rich idle chatter to listen to. Reams, tons, of sensitive material, representing millions of dollars' worth of research and development, are given away by companies in an effort to close a sale: at trade shows, on plant tours and in sales calls. And through research organizations which accumulate massive amounts of information from companies and do thorough competitive analyses, then openly sell their studies for several thousand dollars—peanuts compared to today's research budgets.

I have even heard, and Pooley supported the stories, that into the mid-1970s Japanese businessmen would request, and receive, tours of U.S. semiconductor foundries, and would not be stopped when they asked to take photographs at every step along the way. Some foundries actually turned over schematics of production facilities when requested, they were so flattered that there was that much interest in their business.

The Law—Observance and Breach

One doesn't know whether to be more astonished at the Japanese breach of etiquette or at the American arrogance that thought nothing could threaten our technical lead. And all the while, companies ask their own suppliers and agents—people with an interest in the company's success—to submit to sign-in procedures, escorts while on company property and occasional searches of briefcases.

●   ●   ●

As incongruous as the notion of a gumshoe, in the tradition of Dashiell Hammett, might seem in this high-tech wonderland, the fact is that there are probably fewer places on earth where the need for private investigation is as great.

Investigation? A copy of an intricate circuit design can easily fit under the false bottom of a briefcase. A thousand chips that drive exotic complex devices can be hidden in the lining of a suitcoat, which can be sent out for dry cleaning in East Berlin, Moscow or Bejing.

Private? In the wake of California's budget-slashing Proposition 13, law enforcement agencies are running very lean; the average detective is intimidated (as are most people) by the technology; and the victims—the companies—are extremely reluctant to publicize any slipups in their security.

The offices of McDiarmid and Brown, on Capri Drive in the hillside community of Los Gatos, can honestly be described as nondescript.

There is nothing immediately obvious in the three-room first-floor suite of the small office building that indicates the nature of the work conducted therein. The business card on the secretary's desk shows an embossed eagle with olive branch and arrows. Soldiers of fortune? Quasi-governmental contractors of some sort? The inscription says simply: "Consulting and Investigations." The bookshelf behind McDiarmid's desk is similarly enigmatic, containing a study on body language and some works by Robert Ludlum and Gordon Liddy.

*   *   *

The Law—Observance and Breach

Bob McDiarmid is very low-key, very soft-spoken and very experienced in the art of discretion. Considering the private nature of the work he does, I found him remarkably open and forthcoming in discussing it with me at the time. Only later did I realize just how guarded his replies really were. What seemed comprehensive candor in person appeared more circumspect in playback.

McDiarmid came to private investigation work from law enforcement, a career that started nearly twenty years ago when he was a college student in a small town in Iowa. "I was snowed in in a police station one night and I saw them advertising in a magazine for deputy sheriff for Santa Clara County. I said, that sounds like the place for me, so I came here."

He started as a patrolman and worked up to lieutenant in the organized crime and criminal information section. It was while in that position that he was approached several times by electronics companies, requesting investigations. Though none of those companies was offically located in his jurisdiction—as, for example, in the case where a municipality and not the county would be the appropriate law enforcement entity— McDiarmid found that often the proper agency had no interest in pursuing high-tech criminal investigations.

"I talked to an officer in Palo Alto one time and I asked him how he tracked these cases. He said, 'I follow them to the city limits. When it's outside, that's the end of it.' Now, most cities don't look at other investigations that way. I'm sure if you robbed a bank in Palo Alto and you went to Sunnyvale, they would come down to Sunnyvale and have a chat with you. But it wasn't occurring with electronics thefts. It was just kind of an enigma. Nobody had ever done it and didn't want to do it. They had homicide classes, robbery classes, evidence school . . . but this was a whole new ballgame."

Eventually McDiarmid and some colleagues did get involved. His first case in the late '70s, concerned an employee of a

Valley firm who wished to trade electronic parts he had stolen from his employer for cocaine. The sting operation was successful and "not as complicated as we first thought it would be. Thereafter, we got involved in other investigations. Our biggest problem was the terminology."

In many ways, McDiarmid says, working on the trafficking of electronics parts is similar to working on cases involving narcotics: it's a cash business, with people often selling out of the trunks of their cars. The biggest difference is that the mere possession of narcotics is illegal, whereas finding an integrated circuit on a person doesn't necessarily indicate a crime.

For their part, law enforcement agencies that did get involved in early cases of electronics theft soon lost interest when, after finally recovering the stolen parts, they would find the company didn't wish to prosecute, for fear of having negative publicity appear before investors, competitors and customers.

What happens to hot chips, however, once they do get on the market? A typical case, McDiarmid tells me, would be one in which someone walks off with, or buys, a quantity of unmarked semiconductors, marks them and sells them to a so-called gray-market company—a legitimate vendor of electronic parts who doesn't always inquire where the components he sells come from—in, say, Southern California. From there, they might go to a similar gray company on, say, the East Coast, and from there to, say, Germany, where they might make their way to, say, the Eastern Bloc, and from there to, say, the Soviet Union. In a hypothetical case.

A gray company may or may not know the merchandise was of questionable parentage—the primary concern is that it has suitable-looking documentation. But since these companies also deal in legitimate products, they operate in the open. And they serve a real purpose for manufacturing companies which are occasionally forced to go on "fishing expeditions"

The Law—Observance and Breach

to get components that are in short supply through normal distribution channels.

In spite of the fact that stolen electronic designs or components are shipped under cover to countries whose best interests don't necessarily parallel ours, terms like "treason" rarely come up.

"I think that people sometimes convince themselves that one individual part in and of itself is just not going to be that critical to this country. And if you have quite a few people thinking that way, then you can get just about whatever you want."

So, if the money is good, and the hours short, and the concept of treason now passé, why don't more people traffic in the transfer of chips to other countries?

"There are still some real patriots out there that have a very strong concern about the survival of our system. Because if you turn that around, there aren't going to be too many entrepreneurs under another type of system."

Cases of chip thefts didn't garner much public attention for a long time. Then two things happened that changed all that. One was the theft of $3.2 million worth of semiconductors in November 1981 from Monolithic Memories of Santa Clara; the semiconductors were recovered the following summer at Lake Tahoe. McDiarmid had left law enforcement work by then and represented the insurance company in the case.

Considering the complexity and the amount of product stolen, the theft was pretty straightforward.

"Somebody got to a security guard, and during his shift he deactivated some alarm devices and they rolled in there with vans, and I think he even assisted in loading up the vans. They hauled the stuff off. It was just about that complicated."

When the stakes get that high, it isn't just the trade press that pays attention. And when the stakes get that high, as in the second precedent-breaking case, life sometimes gets cheap.

After one sting operation, overseen by McDiarmid while he was with the county, the owner of a electronics company was investigated by the district attorney's office.

"One of the things we found was that the man had receipts from almost fifty companies for the product that he had in his possession. None of the companies existed, but it made a paper trail and covered him with the IRS. The district attorney's investigator uncovered the individual who wrote up those receipts and interviewed him, and apparently he agreed to testify against his former boss. Shortly before the trial, however, he was bailed out of Santa Clara County jail and disappeared. Two weeks later his body was found, buried in a shallow grave on Skyline Boulevard. No one ever was able to say specifically who caused his death. As it was, the man was convicted without that individual's testimony."

Some months after my meeting with McDiarmid, Larry Lowery, president of a Fremont electronics parts firm, was arrested and accused of the murder-for-hire of David Henry Roberts, whose body was found with several stab wounds and a fractured skull on Skyline Boulevard in September 1981.[16]

Those cases changed a lot of people's attitudes about the seriousness of these kinds of crimes.

"Most of the people that we dealt with had never been arrested, had above-average education, were articulate . . . they were not the type of people that judges are inclined to send to state prison. I think we've seen that turn around now, and the murder probably was the first case where any witnesses really got hurt. I think that changed a lot of judges' attitudes."

Proposition 13 inflicted a pinch on governmental bodies in California, including law enforcement agencies. That plus the fact that there were some areas which law enforcement wasn't attending to, and the old jurisdictional bugaboo—not to mention the entrepreneurial buzz of Silicon Valley—all compelled McDiarmid to start thinking of forming his own private inves-

The Law—Observance and Breach

tigation firm. So a few years ago, he and a partner from the Santa Clara County Sheriff's Department named Wayne Brown, started McDiarmid and Brown.

"Wayne worked for me for two and a half years. He was the first investigator to get involved in electronics cases, and carried the primary load as far as the undercover things that we did, relative to buying stolen parts or selling parts to suspected fences. Previously, he had been a narcotics officer, so he had that ability to get to know people quickly, to establish relationships rapidly and to do deals. He was just a natural.''

The firm does both consulting and investigating. Consulting includes "survey situations, where a company may contact us and say, 'We've got some parts that we have sold to Company B. We'd like you to roll in there and try to pick them up without proper identification, to check our security.'' In addition to designing and testing security measures, McDiarmid's company also consults on safety aspects.

"We had one situation where we had a guy working dock out there. There's a lot of speed being used, and these guys were running around at about a hundred miles an hour on their forklifts. You know, it was just a matter of time until somebody stepped in front of one and got killed.''

Even drug abuse in Silicon Valley is geared to production. According to McDiarmid, the drugs of choice, at least those he comes across, are all uppers.

A large part of McDiarmid and Brown's work is investigation. At first McDiarmid found a big difference between being a private, as opposed to duly deputized, investigator.

"There are times when they hire us now to do one particular aspect of an investigation, just one piece of the puzzle. And I know there are a lot more pieces to the puzzle. But when they say 'Thank you,' then that's it. And it frustrated me at first, because I'd always been trained to do a whole investigation. It doesn't bother me anymore.''

One type of investigative project is preparing backgrounds of companies or people for clients. One of McDiarmid's most bizarre assignments involved a man who applied for a job as corporate scientist at Seagate Technology. He got the job on the strength of his résumé, which said that, among other things, he held over three hundred patents; spoke a dozen languages, including Swahili and Eskimo; was science adviser to the United States, Great Britain and West Germany; served as consultant to IBM, Motorola, Intel, Fairchild, GE, RCA and the Defense Department; held patents on most of the major microprocessors; and possessed two Ph.D. degrees, *cum laude*, from ''Oxford University in Cambridge.''

What's so remarkable isn't that he was uncovered, it's that, in a place like this, no one even thought the résumé extraordinary enough to ask any questions before hiring him for $95,000 a year.

''If you tell a big enough story, people believe it. They want to believe it. Companies here have a hard time filling some of those positions.''

Suspicions were soon aroused at Seagate, however, and McDiarmid was hired to do a fast, thorough investigation. He immediately got on the phone to England.

''I talked to the people in Oxford, and they got very upset when I repeated what he had indicated on his résumé, being from 'Oxford in Cambridge.' '' There were other strong indications that all was not as it should be, and before long ''the gentleman was residing in jail'' on charges of grand theft and fraud in connection with some checks he allegedly bounced.[17]

While McDiarmid has never had his life endangered in his work, at least one individual has threatened to break some of his bones.

''I met in Las Vegas with an individual who was involved in some video game copyright violations. I said I wished to

The Law—Observance and Breach

buy some machines from him. He played a real hard guy. He told me, 'If you're a federal agent, I'll bust your knees.' We were sitting at a table in a casino, and he had a box of printed circuit boards. We reached agreement for me to buy them for a certain amount of money. After he gave me his hard-guy routine, I told him to look under the table, because that's where the money was. But his eyes went woooooooowww! I didn't think of it at the time, but he probably thought I had a gun. He got real nice after that. And paranoid, too.

"We went up to the parking ramp, and every time somebody came by, he would take a shot over his shoulder. Then we hit the streets in Las Vegas and he must have been doing ninety-five . . . in case we had a tail.

"He's been adjudicated now. He's a very sharp individual, and I have no doubt he will survive this trauma and may eventually be a very successful businessman."

I asked McDiarmid if he ever saw much evidence of international espionage, aside from the gray trails through the gray companies.

"When we're doing backgrounds we come across people of Russian origin and Eastern Bloc origin quite often. We don't really have a good way of evaluating how they happen to work in the industry. One of the things that really upset the FBI some years ago was when the federal government cut off the foreign student visas for some West Coast universities.

"Those people were readily identifiable as agents of the Soviet Union because that's all they send over here. The FBI was faced with its becoming a little more difficult to identify who the foreign agents were."

If the feds have a hard time now finding out which students are agents, they certainly have no trouble finding the Soviet Consulate on Green Street in San Francisco's Pacific Heights. It's the building whose roof is a forest of extremely sensitive electronic detectors, most of which are pointed south down the Bay. FBI estimates are that one-third of the Russians who work at the consulate are KGB or GRU agents.

So sensitive are the Russian detectors, according to an article in the *San Francisco Examiner*, that a neighbor came out of his house once grumbling under his breath about a consulate car blocking his driveway and within seconds, a consulate employee bolted out of the building to move the car.[18]

Soviet penetration of Silicon Valley has become so notorious that the U.S. State Department has ruled Santa Clara County off-limits to Soviet diplomats, preventing them not only from any clandestine activity but from the cocktail party circuit as well.

"I doubt there is as much going on at the embassy as we have been led to believe. But it's something that if it wasn't concentrated on might become a serious problem."

It would appear that penetrating American technology secrets is a growth industry for much of the rest of the world.

"For a while the big thing was to give tours to Chinese groups. One of the companies that manufactures aircraft, in the late '70s, gave them a guided tour, but prohibited them from taking cameras. What the visitors were really interested in was the chemical composition of a particular metal being made. They gave them a tour of the metal shop, and one of the visitors had special sticky-soled shoes. He just walked through the shavings and collected particles and had them analyzed, so they knew the formula by the contents."

Does McDiarmid ever worry that the time will come when the computers he's protecting, especially as they come to incorporate artificial intelligence or "expert systems," will replace him in the work he does?

"There are certain human elements that I don't think a computer could recognize. That sixth sense, the hunch, the gut feeling."

*   *   *

The Law—Observance and Breach

In late 1983, the Santa Clara County District Attorney's office announced that it was, at last, setting up a technology theft association. It will begin a cooperative effort between the county sheriff and the police departments of Palo Alto, Mountain View, Sunnyvale, Santa Clara, San Jose, Campbell, Gilroy and Milpitas.

• • •

There is another area in which technology and the law interact, and it doesn't have to do with thefts of trade secrets or microcircuits—hard assets—but rather with thefts in which there is really nothing "missing" afterward.

Thefts of this sort occur in three areas: phone phreaking (making long-distance calls without bothering to inform the telephone company's billing office), system hacking (breaking into computer systems over phone lines to snoop, but stopping short of damaging or erasing information), and software piracy (the illegal duplication of proprietary software programs).

Phone phreaks claim that since there is already electricity flowing through idle phone circuits, and since their ethics phavor phreaking during off-peak hours, there really is no theft. The phone company, they say, simply doesn't make a profit because of a glitch it allowed to be designed into the long-lines switching system.

Most system hackers claim they have no interest in doing harm or taking anything. They merely tap into "secure" computer systems, at banks or on military installations, for the sheer intellectual challenge posed by the fact the system was made "secure" in the first place. (Hacking to perform acts of vandalism to electronic records, or to manipulate data for illegal ends, is clearly different from "recreational" hacking.)

Finally, through the wonders of digital technology, software pirates unlock protected software and pull countless "original" copies of the program to sell or give away, depriving the author of his or her royalty. Again, the argument is that nothing is stolen or missing. The original "original" was paid for, and owners of subsequent originals, it's said, probably never would have bought it if they'd had to pay for it. Either way, the contention is that the author isn't really out anything. (Many affluent, otherwise law-abiding Valley residents maintain this belief. According to computer crime authority Donn B. Parker of the research organization SRI International of Menlo Park, it is called the Peninsula Ethic.)

The argument that nothing is missing or stolen in these three cases is specious. But still, electronic digital technology— with its ability to be tapped into and to provide exact duplicates without damaging the original—has certainly thrown a curve at traditional notions of theft and vandalism.

The generic granddaddy of all these electronic crimes is phone phreaking, which got started in the 1960s when a very thorough researcher, with a less than conventional notion of right and wrong, came upon an obscure technical journal buried deep in the stacks of an engineering library.

In the article, a Bell engineer mentioned, as an aside, the six master frequencies which are combined to make the twelve tones used in all direct dialing. It turns out that AT&T made an irrevocable decision years ago to control all the switching in its long-lines system with those dozen tones made by a proprietary combination of two frequencies each.

When the knowledge gained in that research project got out to a certain subterranean culture, it found practical application in "blue boxes"—multifrequency oscillators that allowed the user to recreate those audible tones at home or at pay phones— permitting long-distance interconnections while leaving no record of the call.

Others, in the meantime, found other ways to create the same effect; one man could actually whistle with perfect pitch to recreate the exact tones.

The whole craft took on a new élan when a Bay Area native made a most remarkable discovery: the 2,600-cycle tone created by a toy whistle found in each and every box of Cap'n Crunch cereal could be blown into the mouthpiece immediately after dialing any long-distance number, thereby

The Law—Observance and Breach

terminating the call as far as the phone system knew, while permitting the connection to remain open. (That loophole has since been closed.)

For this accomplishment, John Draper soon picked up the handle of Cap'n Crunch. In the nether world of phone phreaks, especially in those iconoclast-loving days of the late '60s, the Cap'n soon became a legend. (After all, word travels fast among people who can make free long-distance calls.)

The Cap'n's exploits defy the imagination of those who consider the phone as nothing but a mundane household appliance that hangs on the wall. For the Cap'n and other phreaks, it was the entree to a wonderland that takes up where Lewis Carroll's imagination left off; a system that could take them to the ends of the earth—Moscow, Pretoria, Vancouver—and back again, in a matter of seconds; they could choose the routing of the call—by cable or satellite— and sometimes stack the call on top of itself.

A marvelous account of these exploits, called "Secrets of the Little Blue Box," by Ron Rosenbaum, appeared in the October 1971 issue of *Esquire*. It describes how the Cap'n would call himself—choosing to route the connection through Tokyo, India, Greece, South Africa, South America, London, New York and California—to make a second phone next to him ring. He'd have a wonderful time talking with himself, given the round-the-world delay of twenty seconds. On other occasions, to make things more interesting, he'd use four phones, calling himself both east to west and west to east around the planet.

And what would he say to himself after all that effort? "Hello hello test one two three." Clearly, it was the medium, not the message, the Cap'n loved. Truly loved. Rosenbaum indicates the Cap'n had a passion and a curiosity about the phone system that bordered on the erotic. And there was nothing illicit about the affair, for the Cap'n. To him, the system was an intricate, beautiful creature. He only wanted to explore it. Like a lover, yes, but also like a scholar.

The Law—Observance and Breach

And, as with scholarship, that meant field trips. One can't always just let one's fingers do the walking. According to Rosenbaum, the Cap'n would pull his VW van, loaded with electronic gear, up to an isolated phone booth in the countryside and sit, sometimes for days, calling around the world, sending his voice into that complex network of electronic circuitry and determining where in the world it would surface.

By the early '70s, the Cap'n had developed such a protective regard for the system that he began to worry that unscrupulous users might manipulate the few flaws it contained, to create untold havoc. In the *Esquire* article, he indicates it would be possible for as few as three "radical underground," working at different phones, to "busy out" the entire Bell System. If such a thing was possible (the phone company said it wasn't, but other phreaks thought it might be), the Cap'n would have found no satisfaction. He cared too much for the genius, the complexity, the simplicity of the phone system.

In time, his growing reputation and the demands of being an underground celebrity took their toll. AT&T was not about to let this sort of activity go on, difficult as it was to trace the calls of the better phreaks. Where might it end?

Where indeed? Among those who took up in the Cap'n's footsteps, after reading the *Esquire* article, was Stephen Wozniak, who would later cofound Apple. Interviewed in a subsequent *Esquire* article, "Secrets of the Software Pirates" (by Lee Gomes in the January 1982 issue), Wozniak claims he and partner Steve Jobs began, long before Apple, to build and sell their own version of the blue box in the dorms at Berkeley. Woz became an accomplished user, claiming there is a tape of himself waking up the Pope claiming to be Henry Kissinger. Each Wozniak/Jobs unit came with this inscription inside: "He's got the whole world in his hands."

Less experienced practitioners got into the field, however, and began to get picked up. (Phreaking through an 800

The Law—Observance and Breach

number leaves a trail, and who can honestly claim he was talking with a car rental office in Boise for twelve hours?)

The skein began to unravel for some of the fraternity, and eventually even for the Cap'n himself when others would make calls around the world and terminate the calls into his phone number. Similar activities by others, plus the notoriety that came with the first *Esquire* story, soon sent the authorities after John Draper. In the early 1970s, he was arrested for the first of four times.

Today he drives a Mercedes, vacations in the Galapagos Islands, and is the president of his own successful company, Cap'n Software. He is also the author of one of the best-selling software packages on the market. Draper has undergone quite a transformation from Cap'n Crunch to company president. And, likewise, what started out as a prank done by a handful of very sophisticated kids has grown into a multimillion-dollar scam carried out today by people who use the phone for credit card thefts, Sprint fraud and accessing secured computers over local or long-distance lines. The Cap'n claims his role in that transformation was significant, but unwitting.

I met John T. Draper and his business partner, Matthew W. McIntosh, at a coffee house just off Telegraph Avenue in Berkeley where Draper prefers to outline his programming efforts these days.

Berkeley and Stanford are both, justly, recognized as two of the most important intellectual centers in the world today. And if one has never visited either institution, it is perhaps understandable how their reputations might overlap given their geographic proximity. To distinguish the two for people who have never seen either, I would point out that spray-painted messages like "U.S. Get out of California" are not uncommon on the interior and exterior walls of craft and coffee shops in Berkeley. They are rather less commonly seen on the facades of elegant department stores in the manicured shopping centers close to Stanford.

Draper was clad in orange T-shirt and black pants, and wolfing down a midafternoon sandwich in the coffee shop. His bearded partner was also in jeans and T-shirt, but even so, he projected the impression of the business manager he is.

From the coffee shop we ambled the several blocks to the University of California and found a small, sunlit grassy patch on which to recline, not far from Sather Gate. Draper wanted to take the afternoon sun and air. He claims he never had any involvement in illegal use of the phone system. Rather, ''I was just intrigued with the system and I was experimenting around with the internal codes more than anything else, to map out the system.

''I was discreet about it, but that ended when these blind kids told everybody. Eventually a whole high school in San Jose found out about it, and from then on discretion was no longer possible. So I kind of laid low and kept away from it as much as I could. But did my own experimentation independently of them.''

Most of the phreaks of those days fit the profile that today describes system hackers. With the exception that a lot of the early phreaks were blind.

''Most of these kids are socially rejected in some form or another. They probably lead a sheltered life-style. I mean, I was that way, too. They don't get a chance to be exposed to normal socialized-type things. Especially blind kids. They were very attracted to the telephone. Because it is audio, not visual, okay? They can communicate with each other. To a blind kid being on the phone is just like being next to someone. They all knew each other from a summer camp for blind kids. That's how it became widespread so quickly.

''The establishment would like to believe that the phone phreaks were antiestablishment, that they were out to screw the system. That wasn't the intent. Phone phreaks were noth-

The Law—Observance and Breach

ing more than people devoted and dedicated to exploring the system and learning everything there was to learn about it, not going out and ripping it off. It was a playground. Why destroy a playground?

"But the phone company and the police were so worried about it that it happened. And had they not been so worried about that, had they backed off . . . What they should have done was hired these people, and it would have suppressed the information."

Draper has faced judges on four occasions. His first conviction in 1972, for fraud by wire, resulted in five years' probation. In 1976, he was sentenced to several months in California's Lompoc. Then in 1978, he was sent to a federal penitentiary in Pennsylvania.

And that, according to him, is when the skills enjoyed by a relative handful of individual explorers and pranksters fell into the hands of the underworld.

"I warned the authorities, 'You put me in jail, you're exposing me to criminals who will find this information very useful. Not that I'm willing to give it away, but there is no way that I am going to resist if they do physical damage to me.'

"But they sent me to prison anyway. I was contacted by the underground people there. At first I gave them some bullshit information. What I didn't know was that this information was being verified and checked on the outside. When this information proved to be false, these people came down on me very heavy. I was almost paralyzed from the hips down because of an injury caused by a fracas that happened in prison. You see, they didn't want the access numbers themselves, they wanted to know how to *find* the access numbers. They were smart.

"As a result of that, the phone company has now got more fraud problems. It's increased now. By putting away one Cap'n Crunch, they have created hundreds and thousands of

Cap'n Crunches all over the country. People have developed the techniques that I had developed, and refined them to levels beyond anything I could do. If they had just backed off a little bit and hired me instead, that information wouldn't have been blown beyond proportion."

Finally, in 1979 in San Jose, Draper was sentenced again for phreaking, this time to a work-furlough program whereby he would spend nights in the county jail and work days at a Berkeley computer firm.

At the prodding of friends, he began turning his considerable talents at systems analysis away from the phone and toward computer programming. Having time to write at night, and having access to a computer by day, he wrote a word-processing program for the Apple computer. He demonstrated it at the 1979 West Coast Computer Faire in San Francisco, and a distributor saw it there and offered to market it for him.

"I had a lot of sympathy from the computing world because of what happened. A lot of people were really on my side trying to get me so-called rehabilitated. I owed the obligation to myself and the industry."

The program, called Easy Writer, has proved to be one of the most successful microcomputer software packages ever marketed. At about the same time, he created Cap'n Software, Inc. When we met, he was at work on a new package for his company that will allow people with no experience in programming to create programs of their own. Characteristically, he is vague on details while he's working on it, except to say, "It eliminates this black magic of programming and all its voodoo and alchemy."

Draper is one of a very select breed of independent programmers—those who do it all, alone, so the resulting program reflects them, their personality, their own unique circuit wiring. Programmers like that are beyond being professionals, beyond being craftsmen. They are artists.

The Law—Observance and Breach

"Independents can put out programs much, much better than a team of programmers because it's just one individual. It's one mind, one entity, putting that information into the computer in the form of a personal logic system. Several programmers working on a single program tend to make things very complicated. An individual programmer makes things very simple. It has to be simple for him to understand his whole program from beginning to end. A lot of programmers put their own brains and personalities in there."

The former underground inhabitant obviously enjoys his time in the sun, as he basks for a free half hour on our shared sunny triangle of grass, before another business engagement or round of programming. His recent success and lifting of the weight of the law from his back have given him a new lease on life.

"I get into exercising now and things that are not related to computer programming, so I can keep a more well-rounded type of life-style. You can just get so much of computers. And I need to get my mind off of them sometimes."

6 | CIRCLE TWO: THE ARTISTS

16 | John Chowning, Composer

Data sheets will probably never be acclaimed for their piquant sense of whimsy, or technical articles for the emotional catharsis they provide. But the centuries old division between the technical and the humanistic domains has begun to break down, in Silicon Valley and elsewhere. The posting of the bans, you might say, to a possible marriage of engineering and art.

In the hills above Palo Alto, down a eucalyptus-lined lane, stands a most peculiarly designed building. It sweeps across a hilltop, describing an arc of about 150 degrees, and from a distance appears to be a new-as-tomorrow structure of stunning design. Only when one gets closer is it apparent that this weathered old place has seen much better days.

This building, the Center for Computer Research in Music and Acoustics (CCRMA), is part of the music department of Stanford University. What better hole to fall through to enter this mysterious new wonderland where the computer's wizardry and the artist's sensitivity are merging not for the benefit of the computer, but for the user and his or her audience.

John Chowning, the center's director, had told me on the phone that in order for a conversation between us to have any meaning, I had to come up first to one of the monthly evening demonstrations, to hear for myself what computer music sounds like now.

A hand-lettered sign directs me from the parking lot down an outside hallway, past the open doors of the other rooms in this arching structure. The demonstration is full this evening. About thirty people are packed into a dark room, lined on three sides with black drapes. At the front of the room, a young woman named Jan Mattox stands surrounded by keyboards, terminals, speakers and miscellaneous computer parts. The audience of Peninsula residents ranges in age from early teens to late sixties; some have heard about these monthly demos and come up to check one out, others are regulars.

Jan tells us that this center, the largest such in the country, was founded by Stanford in 1975 after more than ten years of pioneering work by Chowning and his colleagues. The purpose was and is to explore the use of the computer in creating music, as well as illusory spatial environments. Such use is dependent upon programming techniques, signal processing and psychoacoustics, which seeks to understand how we process the signals that we hear: the "silence" of a meadow being different from the "silence" of an enclosed room.

For the people who work here—composers, researchers, students, and guest artists from all over the world such as Pierre Boulez—the computer is the ultimate musical instrument, for it has the ability to generate any sound wave that can be imagined, as well as many that can't. It can also "process" naturally occurring sounds, permitting them to have new textures. The study of psychoacoustics yields information that can give sounds a "personality" or "signature."

The demo begins with the flick of some buttons. And somewhere down at the microelectronic level, numbers repre-

John Chowning, Composer

senting soundwaves are generated in the bowels of a fleck of silicon.

We are seated between four large speakers. In the dark room there is little visual distraction. As the demonstration begins, we are not listening to beep-beeps or ooo-voo-doo-be-dippps, which many think of as electronic music today. We are listening to sounds (sounds first, music later) that are extremely clear (being digitally produced with no tape hiss), compelling, enrapturing. They seem to move around us in aural space. Or are we moving through the sounds? Then electronically generated natural-like sounds—a human voice and a stringed instrument—do what no sounds in nature can do: they transpose one into the other. The effect is spine-tingling.

Is it just the novelty, or can these sounds really convey an authentic musical experience? With that, more buttons are activated, and a complete musical work is played. This is not an atonal, aharmonic, amelodic assault on the senses. The work has all the evocative richness of a nineteenth-century Romantic score.

Because we patiently put up with the old machines that commanded us not to "fold, spindle or mutilate," we have been rewarded by crossing over some aesthetic barrier. In the hands of a skilled programmer/composer, these machines—though they lack one themselves—can touch something in our souls.

The process, however, is painstakingly slow, we're told. A single movement of the work we're hearing took an entire school term to program/compose. First, the appropriate electronic tones had to be synthesized, then sequenced in a pleasing order: making the sound, first, then the music. Everything comes from scratch, the composer typing in lines of programming code to create the timbre as well as the melody, functioning as instrument builder and performer, as well as composer.

As I left that night and drove back down the dark hillside

into Palo Alto, I wondered if what I'd heard was a new kind of music that was beyond music, a sort of metamusic. But then, wasn't Mozart's work metamusic in his time? And Beethoven's in his? And Stravinsky's in our own century?

Several weeks after the demonstration, I met John Chowning at his office at the center. While he was patient explaining his work, it was clear to both of us that much that lies at the heart of this music—or any music—doesn't lend itself to verbal explanation.

When I asked him to define music—whether acoustic or electronic—he could only say it is "the expression of something that seems to need to be expressed in humankind, just as poetry is. We could talk for a long time about what music is. I don't think we can say much more than that. If we knew what music or the visual arts are, if we could say what the effect or function was by using language, then it would be far easier to express that in language, because it's a far more accessible medium."

Chowning's first instrument, as a child in Wilmington, Delaware, was the violin. Later, he took up percussion instruments, and in the navy, during the Korean conflict, he played a lot of jazz.

After graduating from college, he studied music in Paris in the late '50s. At that time, Cologne and Paris were the centers of the new electronic music and *musique concrête*. At the time, this consisted of synthesizing sounds from electronic signal generators (at Cologne) and electronically transforming natural sounds—whether coming from a musical instrument, a buzz saw or a shattering glass—into new sounds (in Paris).

Chowning later came to Stanford to pursue further studies, and soon thereafter heard some early attempts at using computers to synthesize, or create, actual sound waves (as opposed to using electronic devices like oscillators, amplifiers and filters, and techniques like varying playback speeds, to manipulate naturally occurring sounds). The idea of composing on a

John Chowning, Composer

computer—synthesizing, then sequencing electronic tones—intrigued Chowning, and it wasn't too long before he took his first course in programming.

"The notions of beauty are very different for artist and engineer. For example, there's a certain beauty about a sinusoid, or sound wave, and I understand that. It has to do with trigonometry and right triangles. But it's not a very interesting sound from the point of view of the ear. I would say the artistic notion of beauty is a little more inclusive than the engineering notion. That is, an artist who sees the geometric manifestation of a waveform is able to appreciate that beauty. The engineer probably has a little more trouble understanding what artistic beauty is, or why an artist wants to do certain things which seem to be a perversion of the intended use of a machine.

"But, that's the nature of it . . . the artist is not interested in right workings of machines, but in making art. The first sculptor, for example, didn't use a tool that was manufactured for those purposes. It was most certainly a cast-off spearhead or something. So we who used computers for music in the early years were seen as having some sort of perverse intent. That's not what these machines were intended to do, we were told."

For Chowning, the process of humanizing computers will be a long one, but he is sure that artists/software programmers will continue to find capabilities in the hardware which will astonish electronic engineers.

In talking with Chowning I'm struck by how he and other artists I've known talk about their work in a different manner than do engineers. His remarks are on the order of a working draft that gets polished even as he explains further. Whereas, I've found, when engineers speak, what they say has a much more finished tone to it. First pass, for them, is final draft.

The difference isn't so much in the "what" of what's being said, but in the "how." Engineers assert statements;

artists seem to process them before the listener. Perhaps the word that best explains that difference is "nuance": the lifeblood of art, the bane of engineering. Every performance of a given étude has its own personality. Every 68000 16-bit microprocessor from Motorola damn well better be just like every other.

Computer music, Chowning feels, will be "additive, not exclusive." It will be a new branch of music only, not a replacement, any more than the nineteenth-century symphony orchestra has replaced the eighteenth-century chamber orchestra.

"I don't think the San Francisco Symphony is going to disappear because we have more powerful computers and more digital music devices. Loren Rush, a very fine composer on our staff, had a piece called 'Song and Dance' commissioned by Seiji Ozawa with the San Francisco Symphony. That was for orchestra and quadraphonic tape, and the tape was realized here."

"Karlheinz Stockhausen [a leading composer of the Cologne School of electronic music] asked me once, 'Can you write music at a faster rate with computers?' Well, in fact, I don't think so. It's certainly no faster as far as my own experience is concerned, and maybe a little slower."

One of the principal differences between composing music for the computer, as opposed to composing for traditional orchestras, is the degree of control the artist has over the microstructure of the sound. The electronic composer is his or her own instrument builder. For the traditional composer, for example, the violin comes for nothing. "That is, if it's a good violin and a good performer, then there's a wealth of information that's implanted in that local system of performer and fiddle that the traditional composer's not required to specify, that is simply a part of the heritage.

"The computer composer has a kind of control which was heretofore denied simply because the traditional composer hadn't the capability or the time to deal with music at that

John Chowning, Composer

level, to build instruments. And that seems, on the surface, to be an inordinately difficult task because there's so much data. Except that computer programs begin to help.''

The curved building we are in previously housed Stanford's artificial intelligence (AI) research facilities. Because Chowning needed a computer to work with from the beginning, he became a ''parasite'' at the laboratory, working on the periphery of that group, using its computers at night and on weekends for his programming/composing. In 1979, the AI lab moved out and down to the main campus; Chowning's group remained. Now the CCRMA, too, is about to leave the heights for the campus, perhaps symbolizing the mainstreaming of computer music.

Though he was a hanger-on at the AI lab for a number of years, using its computers, Chowning sees little connection between his work and the work of those who would write ''expert systems'': software that attempts to distill the essence of a given process (such as discovering hidden oil deposits, or diagnosing a disease) in order to facilitate/automate/elucidate that process.

''There are some who are interested in automatic composition. I'm not. I see that as a more difficult problem even than computer-generated poetry.'' While AI may vastly speed up some routine procedures, to the point of making the process appear intuitive, he doubts that will ever or can ever happen in the artistic domain. Who can even understand ''imagination'' or ''intuition'' or ''inspiration,'' much less codify it?

''Now there's a certain view toward this. That if you have good AI researchers hanging over the shoulder of a creative scientist or artist, the researcher will see things that this person doesn't and formalize these 'expert systems.' But when you're looking at some chemist doing his work, many things are quite well defined. It's different when looking at a poet write down words. There's so much less that's formalized at the basis of a poetic statement than, let's say, in organic chemistry. We can't even make a very precise boundary as to what's poetry and what's not, or what's theater and what's not. . . . John Cage's notion of music is one that includes anything which impinges upon the eardrum. That's all potential music. You see, the notion of art doesn't seem to allow itself to be rigorously defined, because it is always redefining itself.''

Several years ago the CCRMA began giving outdoor concerts in the hills near the center. Within four years, the audiences had grown from about two hundred to over a thousand people at a performance. The assembled would sit in a natural amphitheater between four large loudspeakers. The illusion of sound moving through space was ''astonishing'' to many. The concerts have now moved down to a performance center on the campus, and the crowds continue to grow. Why?

''We are right on the edge of Silicon Valley, and people come who live and work in that world because they hear 'computer music,' expecting trivial beeps and bloops. Not expecting anything that is perceptually rich or has meaningful properties which would allow the listener to say, 'Hey, I've got to sit down and listen to this.'

''And that's what happens. Continually, we get a 'Gee, I never thought computer music would sound like that.' And they come back and bring their friends.

''Silicon Valley needs a soul. Because it's heavy commerce, heavy engineering. And most of what people are doing is done for purposes that are . . . Well, let's imagine you're working on this video game device that some kid's gonna stare at, bumping little bodies around. It's pretty hard to convince yourself, you know, that this technology has any soul. Well, art *is* soul. And in a certain sense what we do becomes the soul of this industry.

''There's no resentment in my feelings. You can't have machines that are used for artistic purposes without having a large commercial interest in them as well.''

John Chowning, Composer

Certainly, no one would have undertaken the expense of creating computers solely as a compositional tool. But has the purpose of software programming reached its popular height in the creation of video games? Chowning isn't so pessimistic.

"There's something potentially there in video games that may not be seen by the critics. Mainly, kids become fascinated with the processes by which these games are created. And, to some extent, will become programmers in order to do something that has more than trivial entertainment value.

"And in so doing, they will contact the whole world of the intellect which would never have been part of their experience . . . namely, high-level language, which represents thousands of man-years of thought about thought, thought about logic which goes back to the origins of thinking.

"To learn to program, for whatever reason, is already an enriching experience. And though I don't think that that's the game companies' purpose, they're going to have to accommodate to the fact that kids are going to want to program their own games. That's already begun.

"A kid learns to program in BASIC and quickly generates a game, finds it much too slow and says, 'Well, how can I speed this up?' He finds that if you learn to program in ASSEMBLY language, you can make all this happen faster. That's the seductive aspect of it, and to good purpose, I think. It's almost as if it happens in spite of the intentions of commerce."

I am not a programmer. I have never written a line of code. And I wonder, as I leave the center, why Chowning made such a big thing about the seductive nature of programming.

Then on my way home from the CCRMA, driving along the Bayshore freeway, I pass Intel's facility just south of the Lawrence Expressway, and I'm struck by one of those "Aha!" notions.

Whether they intended it or not, Intel and Motorola and National Semiconductor and Zilog and all the others have taken their little microprocessor brainchildren and by the handsful thrown them to the wind, to the world, saying, "Go solve problems."

To which Chowning and others are replying, "Yes, and more than that. Create things, too! On our own. In our own way."

Microprocessors are cheap enough, and mass-produceable enough, to become universally available. But until the ability to manipulate that power, by programming, is equally universal, we are no better off for the technology.

As programming becomes more simple, however, and as that ability becomes as widely taught as verbal literacy is supposed to be now, the microprocessor will open up realms beyond imagining, and not just for professional composers.

Consider the parallel in literacy. In the Middle Ages, the ability to read and write was the private preserve of the few in the state and the church, who could thus manipulate mass ignorance to their own ends. With the introduction of the printing press in the mid-1400s, technology made available inexpensive, mass production of printed material. One has to suspect that the subsequent broadening of the base of literate people is as much a cause, as a result, of the increase in the standard of living and personal liberty in Europe after the fifteenth century.

And books are passive. They don't read, they *are* read. Imagine the impact of the world becoming literate with a knowledge tool that is not passive, but active. Not centralized, but decentralized. Not solely in the hands of an elect or an elite, but everywhere. Useable by everyone.

Imagine the impact when the concept of software extends beyond productivity tools for the officeworker, or games aimed at kids, to creativity tools allowing an individual to generate sophisticated, aesthetically pleasing visual and aural experiences at home; when the computer is an extension not only of our heads, but of our hearts as well.

17 | Scott Kim,
Visual Artist

Scott Kim and I played telephone tag for several weeks before we finally made connections, to arrange to talk about the computer as a tool for the visual artist. Like John Chowning, Kim asked me to attend a presentation first before we talked, in this case a lecture he is to give at Stanford to an introductory class in mechanical engineering on the subject of creativity and visual thinking.

Kim's lecture is to be given in a hall named in memory of Fred Terman. The building is a fitting tribute indeed to a man whose eye for detail searched out the secret techniques of Finnish javelin throwers. The engineering that holds the structure together is exposed for all to see, to such an extent that the elevator is glass-topped so the rider can see and appreciate the ins and outs of its ups and downs.

The two-hour-long mechanical engineering class is already in session when I arrive. Kim meets me outside the lecture room, waiting for his turn to be called in, and shows me a copy of his *Inversions: A Catalog of Calligraphic Cartwheels,*

237

which was published in 1981 by Byte Books. He invites me to peruse it before the lecture, so I take it into the hall with me and begin to thumb through it in my seat in the back of the class. At first, it seems to be only a clever demonstration that a good calligrapher can script certain words so they read the same way upside down or in their mirror image. An amusing diversion, maybe, but not much more. For example:

EXAMPLES I, II

Scott Kim, Visual Artists

EXAMPLE III

Copyright © 1981 Scott Kim

I come across names that Kim has played with, names of men who plant mines in minds: Martin Gardner, J.S. Bach, Kurt Gödel, and that notorious visual terrorist M. C. Escher. Maurits Escher, whose single flight of stairs goes forever up. Escher, whose illustrations are both fish and fowl. Then I see this harmless-looking trade paperback carries a foreword by Douglas Hofstadter, author of *Gödel, Escher, Bach: An Eternal Golden Braid. A Metaphorical Fugue on Minds and Machines in the Spirit of Lewis Carroll.* There is more here than first meets the eye.

I read Hofstadter describe how his friend possesses "double-jointedness of the mind." I read further where Kim claims that in his work he wants people to see how he is fooling them. "To understand the mechanism and still be entranced— that to me is the greatest magic."

I look again at the cover. The title is *Inversions*. The author's name is Scott Kim. He has scripted the title in such a way that read upside down, it spells his name, and vice versa:

I notice something else. His name inverts another way, too. In fact, Kim Scott sounds even more plausible.

Kim is about to begin his lecture. He is introduced to the sixty or so freshmen as a doctoral candidate at Stanford

EXAMPLE IV,

whose interdisciplinary field is called Computers and Graphic Design.

He was formerly a teaching assistant in this Visual Thinking class, but now is only an occasional lecturer on the subject of creativity.

He is frail and boyish-looking when he takes the lector position. He suddenly jumps to make a point. Then continues on in his mild-mannered way. He stops to scratch his head: in bewilderment maybe, maybe bemusement. Then again, as

Scott Kim, Visual Artist

suddenly as before, he makes another grand theatrical gesture. Now he is frail and boyish and mild-mannered again.

Kim asks us to imagine four shapes: an uppercase A; a lowercase printed a; a cursive, handwritten a; and a typewritten a. And what do these four distinct shapes have in common? he asks. That's right, they're all a-shaped. Four different forms, all described as a-shaped.

(Try it on a friend. Ask him or her to draw you *the* shape of the letter a. Can you guess which one he'll draw? Now commission an architect to design for you a cursive, lowercase a-frame house, and ask him what different building materials he will use to support the weight of snow that will accumulate on that a-frame house, snow that wouldn't accumulate on an uppercase A-frame structure.)

Not content to undermine our equilibrium, Kim now sets out to make us look silly. Put a pen in each hand, he says. Now put the two points together in the center of a page of blank paper. Good, now proceed to write your name: in the normal manner from left to right with the right hand; then the mirror image in the opposite direction, from right to left, with the left hand.

Not wishing to look like a klutz, and not being registered in the class anyway, I excuse myself from the assignment. I look around the room, preparing to smile condescendingly when the others look up in frustration. But then . . . what the hell? They all look up grinning and laughing. Something's amiss. I try the exercise. There's nothing to it. It's unnerving.

The ice is broken. Kim now asks us to turn to two neighbors and find a way to shake hands with both of them at the same time. Is there no end to this man's perversity?

Three of us work at it for a minute and finally find a way. Great, Scott (Great Scott!) tells us, and we all feel very pleased with ourselves. Then he drops the other shoe. Now find at least two more ways to do it. The party—and I mean party—next to us breaks out laughing. They have discovered a way whereby all six of their hands are symmetrically enfolded.

Later, we stroll across campus to his office in the computer science department. Kim's is the new-age equivalent of the artist's atelier. It's a tiny, tiny workspace that can be gotten to only by walking through a much larger room filled with a number of minicomputers. It is literally too small for us both to sit down in, so he drops off his books and we find a quiet lounge area in the building in which to talk.

He begins by telling me that when he graduated from high school in suburban Los Angeles, his interests were music and mathematics, particularly recreational math: the kind Martin Gardner wrote about in *Scientific American*.

"When I came to Stanford, I started asking around to see who was doing interesting things in recreational mathematics. Well, the math department was not really the place, and they pointed me to computer science.

"Much of the love I had for math really found a much more proper place in computer science. I loved playing with ideas, but I wanted to see them more concretely. I wanted to make things. Unlike mathematics, computer science allows you to do that."

Kim's interests were soon split three ways: between music, math and computer science. "Coming to Stanford there was a flush of excitement, being able to pursue anything you wanted and having excellent people. I went nuts, looking all over the campus."

It was on his trips around the school that he discovered the Visual Thinking course we have just come from. Oddly enough, he never took the class, though he became a teaching assistant in it. And though he claims to have little formal artistic training, the field of visual thinking has since become a big area of interest for him.

"It's not art in the most conventional sense. It's more

Scott Kim, Visual Artist

pictures used in communication. It's close to graphic design. It's anything where you clarify a use by making pictures.

"I didn't have words for that before, but what I loved in mathematics was geometry, things that are spatial, pictorial, and therefore relatively easy to communicate to other people. Geometric puzzles, illusions, those sorts of things.

"That's what the Visual Thinking course is about, teaching the ability to think in pictures. Something that's thought of as unteachable."

According to Kim, it wasn't the Visual Thinking course that began his interest in letters. That "goes all the way back. I find letters fascinating. Trying to write letters in unusual forms."

He started doing his unique style of calligraphy quite by accident, writing out the names of people in his dorm so they read the same backward and forward. When he saw what a hit his works were, he began to read up on letters and lettering.

"What I find fascinating is not just the forms, but the idea behind them. The idea that there's one letter, like the letter A, but it has many different forms. All of these forms express the same idea.

"I enjoy exploring ideas by looking at how they can be varied, how they can be shifted. How far you can take them in a different direction. Seeing all the different varieties is very exciting."

Why the fascination with Escher?

"He pursued, with a degree of precision that is normally characteristic of the scientist, the structuring of complicated visual images. He was completely unrelated to any artistic movement of the time. He realized he wasn't going to be considered as an artist. He found much more understanding among mathematicians and scientists who said, 'There, look, that's the expression of my idea!' A lot of scientists would latch onto his ideas as the expression of their ideas, and they'd be puzzled that he didn't understand what they were doing.''

It comes as a surprise to me to find out that about a third of the images in Kim's book are computer-generated. I say that's a surprise because the sinuous, fluid lines of his cursive style have nothing in common with the hard, straight lines and jagged edges we normally associate with computer graphics today.

The whole field of computer graphics, however, appeals to Kim. In fact, his area of interest, computers and graphic design, is an attempt to bring together the technical orientation of computer graphics and the artistic orientation of graphic design.

''Computer graphics trains you to think in terms of process. How would you build that image? What steps would you go through to create that image? The other thing is that it allows you to think directly in three dimensions. You imagine it all in your head and then, as the last step, you type it in.''

Kim points out that there are a lot of things that graphic artists do that the computer can facilitate, primarily in the graphic equivalent of text editing. Today's designers can thereby work with an image, change any part of it, and not have to redraw the whole thing. ''I think with computer graphics there'll be a much more active sense of participation. You'll look at a boat or a building or something and say, 'I know how to make that!'

''Many things remain to be visualized, and not just as static images. The computer takes the additional step of making these images something you can use. You can point to a part of the image . . . go in closer and closer . . . and find out more about it. Make use of it.

''Home computers really emphasize color and sound and motion, because those are things that most readily engage your attention. If computer graphics becomes widely available on home computers, and if those computers include the

Scott Kim, Visual Artist

ability to produce moving images in a fluid way, then that form of communication would be encouraged."

Universally available tools alone, however, won't make everyone an artist in any upcoming merger of computer graphics and graphic design. "People in computer graphics who are generally from technical backgrounds are rushing in and saying, 'We can produce pictures—this is marvelous.' But they don't have much artistic training in how to use their eyes or what colors will work with each other. I see a very strong need for people who have that sort of process ability in graphic design to come into the computer field. It will happen."

I ask how personal computers can be modified for use by artists.

"I don't expect that the future hybrid artists-programmers will have to learn programming the way it's taught now. Probably the most important thing will be redefining the relationship between people and machines. We're still approaching machines from old metaphors. We understand now that cars are not horseless carriages. But what's the computer? It's not understood now, not on a general level."

Most of us think of computers as typewriters attached to television sets, with a small record-player-type thing off to one side for memory. That packaging of computer power came about because it was an expedient, inexpensive way to let us get information to and from the ultra-tiny processors. That form is not, however, the only shape a computer can take.

"We're omitting a lot of senses like touch. Imagine a computer that looks like a piece of cloth, washable, flexible, warm, nice to touch, and you can communicate with it by talking. That's very different. It changes your whole idea of what this thing is in your world. That's talking about its physical changes." That's talking about, literally, clothing yourself in intelligence.

The specific area he would like to go into in the future is

Scott Kim, Visual Artist

designing computers that facilitate human interaction. "I see a new career there. What's known as an interface designer . . . something like that. There are plenty of people working on many different aspects of this. Alan Kay was a prime mover in the early '70s at Xerox's Palo Alto Research Center, where he worked on a language called Smalltalk which defined a new philosophy for using computers."

In Kay's scheme, computer operations would be much more visually based, in order to make the machines accessible to a much wider audience. Years later, Apple would incorporate that concept in the design of its Lisa and Macintosh computers, which use "icons" to facilitate commands—pointing at a pictogram of a file or a wastebasket, for example, to represent "store" or "erase."

Kay also proposed the Dynabook in the early '70s, a product idea only now reaching fruition in the so-called notebook computer. As he envisioned them, these lap-size portable devices could handle text editing and music composition, and, using a program called Paintbrush, could allow the user to select a brush shape and a color tone and proceed to paint on the display screen. The Dynabook would also hook to a telephone, hence to any other data source. And sell for under $500! In an essay published in 1974, called "Fanatic Life and Symbolic Death Among the Computer Bums," Stewart Brand said the Dynabook could be available in two or three years.[19] The technology and marketing approach have only now caught up with Alan Kay's vision. And bear in mind, Kay saw all of this several years before the microcomputer.

"A good interface is one that is usable by a good range of people," according to Kim. "It adapts to their needs as they're using it. It becomes much more conversational than a one-way communication. So you're modeling how the computer's working, and the computer's modeling how you're working. That's not really a thing that's well understood yet, though there are people working on that."

*　　　*　　　*

Scott Kim, Visual Artist

As I drive Kim from the campus to his home in residential Palo Alto, he explains what he sees happening. We shouldn't, he says, be content to let computer people design, for example, educational software. Educators should do that. But today they haven't got the technical competence. Clearly, we've a way to go before being completely at ease with, or best using for human ends, the technology made available by Silicon Valley.

"Computer scientists have set up the tools. It's now up to people in other fields, who have the needs, to define what to do to open the doors. To show what needs to be done."

7 | CIRCLE THREE: THE CRITICS

18 | Ted Smith,
Environmental Activist

It has become a cliché that Silicon Valley represents the end of smokestack industries and polluted skies, of antiquated red brick factories and dangerous and authoritarian working conditions. Yes, the pace here is harried and pressured occasionally, but the living is pretty good.

It's easy to miss some coexisting realities: of houses in San Jose shared by four or five Vietnamese families who work on production lines, struggling collectively to meet one high monthly rent or mortgage; of scores of migrant workers who can be seen from the windows of executive suites, bent over in the few remaining fields, with backaches as old as human toil; of carcinogenic industrial wastes, leaching down through the subsoil to the aquifer which is the single source of water for much of the Valley.

Many a tekkie fast-tracker lives a life very much insulated from more crusty, gritty realities. Those who see and speak with only their own kind here—and that is a very easy thing to do in a freeway society—have the validity of an exclusive existence reinforced, to the exclusion of other realities. They risk losing sight of the fact that, for many—not in the Third World, but here in Santa Clara County—"improvements in computer throughput" are not issues of overwhelming interest.

The underbelly of Silicon Valley—the life some lead as others succeed—is hidden from freeway commuters by the eucalyptus groves, the bottlebrush plants and the oleander bushes.

However far-reaching the social consequences of the Valley's technology are and will become around the world, Silicon Valley is a commercial enterprise. And wherever great fortunes are being made quickly, a tender social conscience is usually a rare commodity.

There are people here, however, who do focus on social and environmental concerns. Not because they wish to be burrs under the saddle, but simply because they can't *not* have these concerns. Their type may not spring to mind when one thinks of Silicon Valley, but they are no less a part of this community.

Ted Smith's law offices are located not far from the modern governmental center of San Jose and Santa Clara County. They are in an older building, shared by other public-interest groups, that lacks the well-groomed appearance of many nearby office complexes. His chief area of interest is toxic chemical wastes getting into the Valley's water supply.

Smith did his undergraduate studies at Wesleyan University in Connecticut, spent several years in VISTA in the late 1960s, took his law degree from Stanford, and has had a general law practice in San Jose since 1973. While he keeps that practice up, a sizable percentage of his time is now devoted to doing volunteer work as chairman of the Silicon Valley Toxics Coalition. How, I wanted to know, does an attorney get involved with toxic chemicals?

"It stemmed from my experience with workers' compensation. Where I saw the people, both injured on the job and

Ted Smith, Environmental Activist

exposed to chemicals, ending up needing medical and legal help. The original interest came from seeing fairly significant numbers of assembly workers coming down with what seemed to be bizarre, disabling injuries and illnesses. And I began to learn more about the electronics industry and the fact that it is a chemical-handling industry.''

What really sparked Smith's, and much of Silicon Valley's, interest in the toxic chemical issue was the revelation in January 1982 that a drinking-water well in South San Jose had been poisoned by leaks of a cleaning solvent called 1,1,1 trichloroethane from a nearby underground storage tank owned by Fairchild Camera and Instrument.

In June of that year, some materials stored at IBM's South San Jose facility exploded, injuring 14 people.[20] Clearly, the issue of proper chemical handling and storage was a real one; it was not affecting "gray companies" as much as blue-chip corporations, and it wasn't going to go away.

Subsequent tests showed that other companies, a virtual who's who of Valley semiconductor manufacturers and others, also had leaking underground waste-holding tanks where discarded chemicals were being stored temporarily for later carting to Class I dumps elsewhere. In fact, according to Smith, when the Regional Water Quality Control Board began to examine the extent of the problem, they found 85 percent of the underground holding tanks tested had evidence of contamination nearby, leaks of their toxic chemicals into the subsoil.

It can take a long time for those liquid chemicals to percolate down to the underground aquifer from which comes the Valley's drinking and irrigation water. But there is nowhere else for them to go. And even if all the leaks were stopped now and forevermore, the eventual extent of the damage is still yet to be seen. What has turned up so far, however, is cause for alarm. According to the *Bulletin of the Santa Clara County Medical Society*, thirty-one families living in the vicinity of the leak reported birth defects, miscarriages or still-

Ted Smith, Environmental Activist

births in the three years before the leak was discovered.[21] Was there a cause and effect connection? A study by the State Health Department is currently underway to determine if there is any relationship.

"The incidents in South San Jose, where this has been most carefully looked at, do tend to show a significant occurrence of heart defects in newborn and small children. That's probably the most striking correlation that has been found so far. There are also several incidents of rare forms of cancers that have developed in fairly young children. I think that's one of the scariest things." In fact, over three hundred families in the area are suing Fairchild over the situation.

Possible contamination of the aquifer is a special cause for concern in the South Bay. Unlike San Francisco, which gets its water from Sierra snowfields across the state via the Hetch Hetchy aqueduct, the Santa Clara Valley gets 60 percent of its water from under its own feet.

If poisoned drinking sources weren't enough, local firefighters have an even more immediate concern that caused them to get involved in the issue, since they are the ones who are first called in whenever there is a detected leak, spill, explosion or fire. They wanted the right to know what it was they were dealing with in any given situation.

After the Fairchild leak, the Santa Clara County Fire Chiefs' Association started to put together what it hoped would be a model ordinance regarding the storage of chemicals. The association set to work with a coalition of business and trade groups, called the Industry Environmental Coordinating Committee, and, in the spring of 1982, put forward a first-draft proposal for a Hazardous Materials Model Code to present to city councils and town governments in the county, for adoption as law within the particular municipalities. According to Smith, however, that fire chiefs industry draft had one glaring point that demanded citizen involvement.

"It proposed reporting the types and quantities of materials stored to the fire departments, but not reporting that to the

Ted Smith, Environmental Activist

community at large. In fact, it *criminalized* the disclosure of that information to the community.

"They claimed that they didn't want to have to expose any trade secrets. That was shown to be a somewhat overly broad concern, to put it lightly. It was that point, as much as anything else, that got a relatively small group of people I had known and worked with concerned enough to come forward to say that we thought it was wrong. We thought that there needed to be full public disclosure, with exceptions for legitimate trade secrets. The issue had gotten so far out of hand that we needed to have a public that was as informed and knowledgeable as possible, so the people could make decisions in their own lives about where they wanted to live, where they wanted to send their kids to school, where they wanted to work. One of the factors that seemed to us to be important to making those decisions was the ability at least to find out what kind of chemicals were being stored and in which areas."

Thus was born the Silicon Valley Toxics Coalition, with involvement of labor groups, homeowners, firefighters, clergy, victims of industrial waste exposure, and representatives of the medical establishment and the environmental movement. Their initial program was to counterbalance what they saw as the enormous strength of the industry coalition.

While the group was originally formed around the right of the public to know which wastes were where, it soon got involved in other aspects of the proposed model code as well: providing protection against leaks by requiring double containment of holding tanks, electronic monitoring of underground tanks, mandatory reporting of all leaks and spills, protection for "whistleblowing" employees, and industry financing of compliance as opposed to a general tax liability.

A revised code containing these items, some of which were put forward by the coalition, along with an expanded list of covered substances which the coalition also pressed for, was subsequently passed by governmental bodies throughout the county. The results?

"Most of the companies have been pretty cooperative with the Regional Water Board in the sense of responding to requests for information, and of hiring consultants to do the testing that has to be done.

"The effectiveness of the cleanups is open to some question. The guy who has been hired by San Jose to implement the ordinance here was quoted in the media the other day as saying that something like only five to ten percent of the stuff that has been dumped into our ground and groundwater has been successfully taken care of. Now the companies want to cut back on cleanup operations." Whatever the percentage, the fact remains that wastes are difficult to remove, and its not at all known yet how much can even be removed.

The problem is enormous in scale. Cleaning an underground spill, never mind a whole aquifer, is a major engineering feat. According to Smith, Fairchild's cleaning operation has been "extensive." It is estimated by the Regional Water Board to have cost the company $10 to $15 million so far.

"Water quality experts are extremely concerned that it may be too late for our water supply here. I think only time is going to tell that. If we lose our water basin, we're going to have to buy water elsewhere. Maybe we'll get into a fight with San Francisco [over Hetch Hetchy].

"Water wars have been a pretty important part of California history. Of Western United States history. And if we lose our aquifer, there's gonna be a hell of a one on our hands."

The coalition feels that while the aquifer is localized, the problems posed are by no means local ones. And though the group started out as a community citizens' group, it may have a larger role to play.

"Those of us who have been through it here feel that whether or not we can save our aquifer, what we've got to do also is to help inform others elsewhere. Particularly with high

Ted Smith, Environmental Activist

technology moving to other parts of the country and the world. The promise, the potential of high tech really has to be balanced against the hazards and the dangers. And people need to be informed of what they're getting into when they buy into the high-tech revolution, at least in areas that drink out of the groundwater. I think that's one of our important jobs.''

Smith finds it curious that some companies that have had the biggest problems are foreign-owned, as Fairchild has been since 1979, when it was bought by the French firm of Schlumberger.

"I wonder whether the fact that they do seem to have such serious problems has anything to do with the fact of absentee ownership. The people who run most of these companies live here in this valley, although most of your top managers don't live on the valley floor. Most of them live up in the hills and up in the northern part of the county . . . the air is a little better up there.

"They may not be drinking the same water that the valley-floor people are, but at least they're here. Not only that, they may be personally concerned about the health of their families. And they're here subject to public scrutiny and pressure.''

In my own work I have come across two distinct approaches to the matter of industrial pollution. Even before meeting Smith, I'd heard that one of my clients was on the list of offending companies with toxic leaks. I called my contact at the company, a midlevel manager, and told him the bad news. He checked around and called me back, and informed me that yes, the company knew it was on the list, but not to worry. The offender was really a firm nearby, coincidentally engaged in the same business, using the same chemicals. My client was sure the leak was the neighbor's responsibility. End of consideration of the subject. Not just with me, I gathered, but within the company, as well.

On the other hand, I also represented another company that, before the ink was dry on the articles of incorporation, spent a quarter of a million scarce start-up dollars to create a model waste treatment system within the plant, one so effective it filtered out the impurities and left them in the form of a solid, dry cake, which is much easier to dispose of than more bulky liquids. The treated water was returned to the city, cleaner than when it came into the plant. The company founder told me that his reasons for incurring the expense were simple: ''Anybody who lives in the environment is an environmentalist.''

Smith feels there's some cause for optimism. ''I think at least the major companies are being more careful now. We've all learned a good deal going through this whole process. And I think everybody is well aware of the potentially enormous liability for doing things wrong. There's certainly a larger incentive as well as greater understanding.''

His chief remaining concern, of a controllable issue, is whether the materials used in much of the electronics industry are so dangerous that there may not be any safe way to store and handle them at all. And that gets at the classic conflict between private industry and the public welfare. On whom does the burden of proof rest regarding the safety of a given industrial chemical's use and disposition? There are over seventy thousand chemicals available to general industry today, with a thousand new ones introduced each year. Where is the balance between progress and safety to be struck?

''Industry would say the general community has the burden of proof showing that something is clearly unsafe before they will make any change. I believe it should be just the opposite. Once there is a sufficiently minimal level of indication that the substance has serious health effects, the burden should be on the industry to show that it's safe. And if they can't show that, then it should either be banned or used only under

Ted Smith, Environmental Activist

extremely limited circumstances. That's going to be a big battle. It's going to be coming down for a long time."

The social issues confronting Silicon Valley, according to Smith, go far beyond a polluted aquifer.

"This used to be called the Valley of the Heart's Delight. That's mostly gone now. We have become a major metropolitan area, major manufacturing area, major industrial area. And we've suffered as well as benefited.

"Take the layout of production facilities. When you do a changeover, you're under constant pressure, you've got to do it fast. You're losing money for every hour you delay. The wiring, for instance, has to be redone. Sometimes it's redone sloppily.

"Pipes carry deadly gases . . . arsine, phosphine, incredibly nasty stuff. A contractor was telling me that when he has to route a pipe, he often finds the original layout and scheme and design are long gone. So when he comes in and has to do a repair job, he can't even begin to figure out what's what. Just real mundane problems like that have been the casualties of the rapid growth."

If facilities have sometimes suffered in the rush, so, too, have some people.

"There are human casualties of the process, whether it's Atari laying off nearly two thousand people with little or no notice, or the day-to-day, grinding routine of the production jobs.

"This is certainly one of the wealthiest areas in the world. And the public school my first-grader went to last year was closed due to lack of sufficient money."

What more poignant way to symbolize the paradox of this valley, where only one out of three families has school-age children? In the fall of 1983, the San Jose Unified School District became the first American public school system in forty years to go bankrupt—this in one of the five richest counties in the country.

Ted Smith, Environmental Activist

Many local Hispanics were here picking fruit long before the better-educated newcomers arrived, but they missed the new gold rush. Something like 35 percent of them today live below the poverty line. And with many of them being unskilled and without fluency in English, what chance have they for that much-vaunted self-improvement when the public school system is bankrupt?

If there is a certain I've-got-mine syndrome here, it's not restricted to the business community alone but infects the larger community as well. In a recent four-year period the city of San Jose added 114 police officers to the force, while laying off almost exactly that number of parks and library staff members.[22]

"I think a lot of people have gotten too rich too quickly, and don't have any sense of how to use that money, or have any sense of responsibility to anybody else.

"I often think one of the nice things here is that there is not an entrenched establishment. Everything is so new that everybody is new to it, and people haven't had time to get so entrenched that you can't maneuver. But there are some stabilizing influences that can go along with entrenched power. Maybe things are a little rougher here because of a lack of that."

Smith's interest was in discussing the project he is involved with now. Regarding his own life and background, and why he pursues public-interest work here in the Valley of the Entrepreneur's Delight, he spoke little. But the little he spoke, spoke volumes.

"I was certainly shaped by the '60s. I think I trace much of my present attitudes to my experiences in VISTA in the late '60s.

"I was in Washington, D.C., where about once a week we'd see a major demonstration or riot. That's when Martin Luther King was killed, April of '68. The whole town was in

Ted Smith, Environmental Activist

flames. That was when the Great Society came to a crashing close.

"I think . . . I think I got seared as a VISTA. I think things were so intense then that I just haven't been able to put it out of my soul. Maybe that's the difference.

"I think I've also been fortunate to be around people who are mutually supportive."

Joel Yudken:
Swords into Plowshares

If there was once an Iron Age, this surely must be the Age of Irony. And nothing highlights this more dramatically than the silicon chip. Alone of all our tools, it extends both our minds and our muscles. It represents at once both a magic carpet to the fulfillment of many of our fondest wishes and the way to efficient, economical annihilation.

Pure science goes about its work without values. It examines what is and finds what it will; it lets the chips fall where they may. There is no such thing as pious chemistry or charitable physics.

Technology—the application of science to immediate problems—has always been regarded as a poor stepchild to science. It is, after all, not pure but applied, not impartial but commercial.

But technology, unlike science, admits the issue of values precisely because it entails application. The decision to use a hammer to build a house rather than to smash open thy

CIRCLE THREE: THE CRITICS 263

Joel Yudken: Swords into Plowshares

neighbor's head to get at thy neighbor's wife is an application of values to technology. So, too, is the decision to use a microprocessor in a pacemaker rather than a missile.

One confronts on one hand the seemingly boundless good that can come from semiconductor technology: in improved crop yield through timely soil analysis and weather prediction; better medical treatment made possible with faster diagnoses, even over long distances; more rewarding learning tools, offering interactive, self-paced and esteem-enhancing instruction. Then one weighs that against the speedy and precise delivery of terrifying new weapons of electronic warfare. The Age of Irony, fittingly, finds expression in Paradox Valley.

Burned by the fickleness of the Defense Department, which in turn was a victim of fickle administrations, many Silicon Valley firms backed off from heavy reliance on the military as a customer in the late 1960s and early '70s. And yet, on two fronts, this place is still heavily involved in the seemingly endless arms race.

First, though they are not often thought of as typical Silicon Valley companies, many local firms are very reliant on direct military contracts. By far the largest employer in Santa Clara County, with over 21,000 employees, is Lockheed Missiles and Space Company of Sunnyvale. (Second place Hewlett-Packard employs about 16,000 locally.)

Lockheed-Sunnyvale, which ranked first among defense contractors in Santa Clara County in 1983, received about $1.5 billion in military contracts that year. Among other projects, the company has been prime contractor for the Polaris, Poseidon and Trident missiles for the U.S. Navy.

Other defense-related firms in Silicon Valley include FMC Corp. of San Jose, with 5,700 local employees, whose Ordnance Division designs and manufactures military vehicles. It did about $1 billion in the county in military business in 1983. Ford Aerospace and Communications of Palo Alto (4,400 local employees), GTE Sylvania of Mountain View

(2,800 employees) and ESL Inc. of Sunnyvale (2,300 employees) are all involved in C^3I—command, control, communications and intelligence—work for the Department of Defense. In 1983, a total of nearly $4 billion in defense contracts were let in Santa Clara County.[23]

In addition to these companies in the Valley whose work is primarily defense-related, a number of leading electronics companies supply components to defense contractors as part of their business, or else produce "ruggedized" computer systems for the DoD.

Nowhere does the matter of values in technology come up so dramatically as with respect to the issue of war and peace. No doubt, most would agree that war—particularly nuclear war—is to be avoided. How to assure the peace is, of course, a matter for debate. Down weapons overnight, even if unilaterally? Phase out the bombs, and verify each other's compliance? Or continue the arms race, seeking "peace through strength"?

Meanwhile, the stockpiling goes on at the rate—in this country alone—of eight new nuclear warheads a day. There is probably no better representation of humankind's intelligence set against itself than in this conundrum: we are knowledgeable enough to build and deploy these sophisticated devices, but not yet wise enough not to build or deploy them.

Of all the endless facts and figures regarding throw weights and megatonnage used by the various parties to support their arguments, the most overlooked, but deeply felt, statistic is those hundreds of thousands of men and women at Lockheed and FMC, at General Dynamics and Rockwell International, at McDonnell-Douglas and the Lawrence Livermore Lab who would be out of work without contracts from the Defense Department.

A 1982 study by the California Office of Economic Policy Planning and Research found that military spending in this state through 1986 "could deplete the entire supply of electrical engineers" from California schools. So pervasive has

Joel Yudken: Swords into Plowshares

defense spending become, so directly dependent are so many people in this country on the arms race and a state of perpetual preparedness, that a compelling argument is made for its continuance by virtue of the fact that we can't afford to stop it. The same problem confronts the Soviet Union.

If technology is to help man the toolmaker promote his continued development rather than threaten his survival, and if the notion of applying values to technology is to have any real-world referent, then surely sound economic alternatives and realistic implementation opportunities must first be proposed to compete with economic arguments for sustaining the arms race on both sides.

The notion of converting swords into plowshares—an idea as old as Biblical times, as relevant as the latest defense appropriation—would seem to be one approach, though at first blush it appears quixotic. In our no-nonsense, bottom-line-oriented world, any notion of such conversion must make economic as well as idealistic sense. But global arms control to minimize the risk of technology-based humanicide begins with personal conversion; with individuals believing such a thing is even possible, in the face of so much momentum in the opposite direction.

Joel Yudken's story is one example of such a change of heart.

"I got a grant from the National Science Foundation while in high school in Queens in the late '50s to study semiconductor technology and all these things way advanced at that time. This was exciting. To be told you're part of an elite, part of a priesthood. You're going to learn the secret language of our society . . . mathematics and science and engineering. And it is heady stuff.

"Later I got into a graduate program which involved working at a corporation and having them pay my education. It's called the Honors Cooperative Program at Stanford. I came out here to be part of that program, and started working at

Lockheed Missiles and Space Company in 1967. And I have to say that the first year that I was here I was very excited, awed, impressed, gung-ho. I was very supportive of our defense policies, including the stance toward Vietnam at the time.

"I identified the SDS as a bunch of crazies. It really was very upsetting to see all that activity on campus, because I was on campus a certain amount of time. Being a pretty straight engineer type I didn't identify with the student movement that much. They weren't addressing a lot of the concerns I had.

"I was in a really good place . . . good job, good school, good future, working on interesting, exciting stuff, part of the defense establishment and in one of the most important, exciting areas in the country. . . . So why like a *meshuggener* did I give it up?"

Joel Yudken was no doubt very clean-cut, wearing a suit and tie and projecting an earnest, no-nonsense look, when he had his photo taken for an ID badge at Lockheed in 1967. Now he's bearded and dressed in a blue chambray shirt and jeans as we talk in a back room near downtown Mountain View, sitting on a few pieces of donated furniture. The sunlit front room is filled with rows of cabinets, containing the reference and research files of the Center for Economic Conversion, of which he is director for programs, as well as of the Pacific Studies Center, a public-interest information center.

"I did something back then which turned out to be pretty fateful for me. I went out of my way to find some reading material that wasn't coming from the radical press, or the antiwar movement. Because I felt that was too shrill, and it didn't appeal to my intellectual and rational engineering mind.

"That somehow burst a bubble in my head. It forced me to think that all the information I've been getting all my life wasn't totally correct, or wasn't the total story. Secondly, I really didn't know a lot at all. So I started to question. I was

Joel Yudken: Swords into Plowshares

still an engineer and working in the defense plant, and then going to campus. I was getting very schizoid."

In 1969, Yudken along with his roommate, other students, and some of the Stanford faculty, helped put together a convocation on Science and Society at Stanford's Memorial Church. It was part of the national "Day of Concern," held on March 4 that year, and first proposed by biologist and Nobel laureate George Wald. Similar convocations were held that same day at MIT and Harvard. Over a thousand people turned out at Stanford.

"We touched a nerve about this thing called technology, and how it was related to the world out there, and that there should be some social responsibility among technical people. The background was the Vietnam War, because this was a very real way that technology was being used, more than ever before, to kill people. At that time, not many of us were thinking about the nuclear question."

Afterward, Yudken was amazed at the number of those who "came out of the woodwork" to share their feelings.

"We were all thinking about the same thing, and it's remarkable the feelings we had, about our jobs, about the environment we were in . . . the sense of somehow wishing we could be doing something else, able to use our technology for something else.

"It's a very hard thing for people who have gone through so many years of training. It's a really grueling period . . . it's demanding work all the time. And you get there and you say, 'What the hell am I doing here? What does this have to do with the dreams I had about what I was supposed to be doing in life . . . that I had when I was younger . . . that got me enthused in the first place? What does this have to do with Sputnik and being part of this great forward thrust of the New Frontier and the John Kennedy era, which got me into the engineering program in the first place?'

"The thing that happened in Vietnam was that we learned

it was a lie, or we felt it was a lie . . . and to some extent it was a damn lie. And that blew our minds.''

It became apparent to Yudken that he could no longer live his split existence.

''I didn't belong there anymore. I had no desire to be there. I hated the job . . . got sick of the environment. I saw a lot of people in a mental graveyard, a mental assembly line. I saw men twenty years older than me, doing work that was just incredibly boring and meaningless, and I knew they were going to be laid off and they had no future.

''I put a peace symbol on a deck of computer cards, and that's what initially drew attention, because somebody in computer operations said, 'Oh, no . . . peacenik! SDS!' Which is incredible when you think about it, but it shows the sensitivity and mentality of the environment. A place like that has a high security orientation, is extremely paranoid. You do the slightest thing and it overreacts. It's like a giant dinosaur that just wags its tail, but you'd better watch out because that tail's pretty dangerous.''

Yudken stopped pursuing his Ph.D. at Stanford and went to live and work in East Palo Alto, the black enclave of the Peninsula, where he took a job as a part-time teacher's aide at a children's center.

''Working with children, especially very poor black children, in an environment which had a lot of violence and a lot of poverty around it was an experience which tested me and required me to reassemble myself.''

He rediscovered a childhood interest in art and began to reconstruct his life as a craftsman, working in leather.

''This was a chance to make a living and do something that was really good and beautiful and people really appreciated. My experience with crafts people was . . . it was a community. It was almost like discovering a basic value and living it for a while.''

Eventually, however, it became apparent to Yudken that the idea of a counterculture, somehow existing separate from

Joel Yudken: Swords into Plowshares

the larger culture, was false. "A counterculture exists within the society and its meaning is counterpoised within the society." Within, not separate from.

"I started to run across the realities that we were in a broader economy, whether I liked it or not. It became harder and harder to keep away from it. When leather prices jumped three or four times because of international markets, you start paying attention and thinking, How come it's so hard to do what I want to do, to do beautiful stuff? You see people doing crap and making a better living, you start to see the operation of the market and how it works. That's when my interest in work life got revitalized."

Soon after that, Yudken reentered Stanford to work on a graduate special program. And beginning with an attempt to write about the crafts movement and its relationship to industrial society, he soon expanded his interests to the larger issue of the overall control of industrial production in society.

In his days in the crafts movement, Yudken had run across the Mid-Peninsula Conversion Project, a group that had been started in Palo Alto in 1975 by Dave McFadden and Natalie Shiras, as a nonprofit research, education and organizing project, concerned with the issue of defense dependency in Santa Clara County. The first organization of its kind in the county, it sought ways to shift capital, technology, skills and resources to the production of non-warfare-related ends.

At the time of his first encounter, Yudken discounted the whole notion, thinking it was impossible. Back in school in 1979, with his interests having moved from an individual's craft work to a study of the larger economy, he was quite prepared to take on the program director's position within the organization when it was offered to him then, all the while keeping in mind the inherent frustration stemming from the fact that in today's interlocked economy, any changes in Santa Clara are ultimately dependent upon changes in the global economy.

Joel Yudken: Swords into Plowshares

The goals of what has since become the Center for Economic Conversion (CEC)—as of similar programs in Boston, Los Angeles, Seattle and elsewhere now—are to promote conversion of industrial facilities from defense-related to non-defense-related work, as a model strategy that will provide a sound economic basis for disarmament.

The program's goals include educating the public about the impact of the arms race on the overall economy, and explaining how economic conversion links efforts to end the arms race with efforts to revitalize the economy.

However quixotic the notion of conversion may at first appear, the concept received some impetus in England in the mid-1970s when shop stewards at Lucas Aerospace, representing about 12,000 workers and seventeen plants, drafted a production plan offering the management of this major British military contractor 120 alternate products the company could shift over to, to avoid upcoming massive layoffs.

Because of several factors, including the shift from a Labour to a Conservative government in the late 1970s, the Lucas Combine failed to achieve its conversion goals. But several other things did follow from the situation, including the formation of a Centre for Alternative Industrial and Technological Systems in the U.K., and the creation by the Greater London Council of a conversion committee of its own in view of London's heavy defense dependence.

For Yudken, there are three main arguments for conversion, aside from the direct threat posed by a nuclear war: military spending has tied up both investment capital and technical know-how, and contributed to the decline of our industrial competitiveness in world markets; defense-dependent communities like Los Angeles, Santa Clara and Orange counties now risk tremendous economic dislocations whenever there are periodic defense cutbacks without adequate alternatives in place; and conversion can break the wasteful cycle of costly budget overruns that are common in the "Iron Triangle," the *ménage à trois* of defense contractors, Pentagon officials and

Joel Yudken: Swords into Plowshares

congressmen (and their Soviet counterparts) who, it is suggested, are the locomotives of the arms race.

These abstractions become a bread-and-butter issue in view of the fact that Santa Clara County is the second-largest (after Los Angeles County) defense-dependent county in the nation's most defense-dependent state. (Nearly 30 percent of all Department of Defense contracts go to California industries, which industries employ over 600,000 people in the state.) Santa Clara County's per capita defense dollar income is $3,250—that's income per person—which places it, according to some estimates, as perhaps the highest in the country.

What's wrong with that, from the county's perspective? It means jobs, doesn't it? Indeed it does. It also acts as a disincentive for any California legislator or businessman to have much interest in any such thing as arms reduction, when there is so much riding on the maintenance of the juggernaut.

This issue lies at the heart of Paradox Valley, for it is safe to assume that Silicon Valley—the headquarters of American high technology—would be among the first targets in any exchange of nuclear weapons. Yet Santa Clara County is so dependent on defense appropriations for communications hardware, missiles and aircraft equipment, computer systems and electronic components that, as things stand, the county's economy could be seriously injured by a significant arms reduction.

Even as other places might wish for such generous government largess, a boom like this actually conceals a number of ticking time bombs. To wit:

● Regarding employment: According to the government's own figures (Bureau of Labor Statistics, 1981), every $1 billion invested in the production of nuclear weapons directly creates 24,000 jobs. By comparison, civilian industry in general creates about 38,000 jobs for every $1 billion invested. Service industries (like teaching) manage to get over 100,000 jobs per $1 billion investment.

● Regarding inflation: While defense-related work puts

nothing into the consumer marketplace, it does pay people to make weapons. There is, therefore, more money in the economy chasing a constant number of consumer goods, bidding prices up the inflationary spiral.

● Regarding capital formation and international competition: Every $1 billion spent on weapons—which at best are useless, and at worst are mega-destructive—is $1 billion not available to the business community to buy and build newer, more productive plants and equipment with which to compete in international markets. According to Seymour Melman, co-chairman of the peace organization SANE, for every $100 of new fixed capital formation available in this country in the late 1970s, another $46 went to the military economy. In Japan, by comparison, less than $4 went to the military. In Japan, therefore, more capital is available for research, manufacturing and marketing of products that compete with U.S. merchandise.

In 1980, according to Melman, Japan's productivity grew by over 6 percent. With our aging industrial facilities, average output per person suffered a 0.5 percent decline that year.[24]

Entire industries in this country, such as steel production, shipbuilding, apparel manufacturing, consumer electronics, automotives, machine tools—areas in which we were once the undisputed leader—have all suffered tremendously because of the unavailability of investment capital for continued modernization.

Not only are funds not available for business investments, or investing in the future (education), but capital isn't available for investing in the so-called infrastructure necessary to any society: the building and repairing of roads, bridges, dams, mass-transit systems. Likewise, mortgage money for homebuyers is in short, expensive supply because of the drain of capital from the consumer to the military sector.

Too, creative resources become tied up by the military, with about one-third of all engineers and scientists in this

Joel Yudken: Swords into Plowshares

country doing work supported by the Defense Deparment, while two-thirds of all federal R and D is defined as military-related.

Clearly, conversion from our tremendous dependency on defense spending, by whatever name or however implemented, touches upon more than arms reduction. It affects our entire industrial and social reinvigoration. It raises the question of whether there will be enough investment capital and skilled engineers left over to distribute the promise of Silicon Valley to other parts of the country. And most estimates suggest that the Soviet Union is similarly afflicted by its heavy investment in arms production, at 7 percent or more of gross domestic product, compared to 1 percent for Japan.[25] Rather than gloat over the fact, we might find that a common meeting ground for mutually beneficial reduction.

"Conversion is an attempt to lay down a basis for providing security and protection for workers and communities in those defense-dependent areas. It allows congresspeople to get off the fence and vote based on real national security, rather than the pork barrel. There are campaigns in their districts led by defense contractors who say if you vote this down you won't get money from our PAC and you won't get the support of our workers because it will cost ten thousand jobs.

"You have this paradoxical situation which I call a military addiction. It's like a drug addiction where you've got to have that fix to keep going, keep on pushing. And yet your overall health is declining. So we have to develop our own 'methadone' program, and the conversion of industries from military to nonmilitary work is part of that. It's a strategy for implementing a methadone program for the society."

Even politicians who may want to work for arms reduction but have defense-dependent constituents are between a rock and a hard place. Senator Alan Cranston of California voted, on one hand, to appropriate funds for the B-1 bomber, and, on the other hand, for the nuclear weapons freeze.

"Unless we have an economics program in place when we do our political negotiations on the arms race, on both sides—because the Soviets have a major problem of their own of the same kind—we'll have many tens of thousands of jobs threatened in the Santa Clara Valley, hundreds of thousands of jobs threatened in California right off. And perhaps millions threatened throughout the country."

Who wants to negotiate massive layoffs?

There is, of course, the risk that conversion is an abstraction, a notion, an ideal, but no more than that. We may soon find out. At the time we spoke, the CEC was just becoming involved in a program at the McDonnell Douglas (MDC) facility in Long Beach, California, that may prove the economic viability of conversion in a very real way.

Though MDC is one of the largest defense contractors in the country, the Long Beach facility is primarily involved in the manufacture of commercial aircraft. It has, however, been plagued with boom-and-bust cycles for over two decades. Whereas the facility once produced the entire aircraft, most sub-assembly work went elsewhere, leaving only final assembly of modified DC-9s and DC-10s for Long Beach. As a result, the workforce was cut by nearly ten thousand during a two-year period, and the plant left running at 30 percent of its capacity in the early 1980s.

Frustrated by this latest chapter in a frustrating situation, the United Auto Workers Local 148 asked the CEC and the California Department of Economic and Business Development to come up with a way to help save the facility and the jobs at Long Beach.

That program, still in its tentative stage has drawn up a list of possible alternate or additional products—such as medical devices, commuter aircraft, light-rail vehicles—which could be made at the existing facility, to use excess capacity and create new jobs for laid-off workers.

The overall criteria for product selection included social

Joel Yudken: Swords into Plowshares

usefulness in the civilian sector, commercial viability, and best use of skills and capacity at the plant.

Having selected and targeted a project that would allow the plant to manufacture light-rail vehicles for an urban mass-transit system, under contract to a prime contractor, labor and management then began talking together, in the face of a history of some past antagonism, about bringing the contract to the plant.

If the program develops, it "would be the first example of employee-led initiative for this kind of new enterprise development in the U.S.," and could show the way to deal with the problem of how to take existing aerospace resources and transfer them to the production of alternative, nonmilitary, commercially successful products.

If the plan succeeds, it could become a model for replication elsewhere, both in defense-related as well as non-defense-related industries, as a realistic route to arms reduction, to worker participation in corporate product planning, and to the development of new technologies.

If the United States and the Soviet Union wish to compete with each other, each trying to prove the superiority of its system, let them compete in economic output, not continuing to use technology to risk the existence of every living thing. Indeed, it might be said that the great debate of our time is no longer between these two competing systems, but between those in both camps who are willing to maintain and increase the technology-based arms race and those who wish to minimize and possibly eradicate the risk inherent in that race, while, at the same time, reinvigorating their stagnating economies by using the new technologies to upgrade their industrial base.

I live several valleys removed from Silicon Valley. On my way home each evening, through one of those valleys, I pass near the Lawrence Livermore National Laboratory, where they design the bombs that are the centerpiece for all the other paraphernalia of the arms race.

Tonight, driving on Interstate 680 through the pass between the Santa Clara and Livermore valleys, I look at the vanishing lights of Silicon Valley in my rear-view mirror. In a small workspace in Palo Alto, Scott Kim is working on ways to foster creativity. In a hillside residence in Cupertino, Catherine Gasich is having dinner, perhaps prepared in an appliance powered by a microprocessor. In a studio in Mountain View, Dick Steinheimer is finishing up photographing a telecommunications device for a product brochure. In a semiconductor foundry, the evening shift is producing chips that will monitor life-support functions and make missiles smart. And in spite of the best efforts of Bob McDiarmid, some of those chips may get stolen and shunted off for use in Soviet missiles aimed at us. The poignancy of that prospect is the irony of our age, the paradox at the center of this place; intelligence set against itself, creating and threatening.

And so, in coming to the farthest out concentric circle, the last orbit, we come to the core of any search for the meaning of Silicon Valley. Ultimately, the significance of this place has less to do with the high-technology tools that man makes than with defining who this toolmaking, thinking animal we call "us" is. However inward-looking Silicon Valley is in many respects, it is, like the rest of us, involved in mankind.

8 | WORLDS BEYOND IMAGINING

20 | John Peers: At a Far Frontier

It is easy to lose sight of how imaginative the engineers, the programmers and the tinkerers of Silicon Valley have been, taking private fantasies that had no precedent, making them part of the fabric of daily life throughout the industrial world.

Their far-out became our standard operating procedure. In order not to lose sight of how remarkable their accomplishments were, it's revealing to look at the visions that are being pursued here now, today. It will give some clue as to how outrageous dreams sound while they're still dreams.

It is seldom that really significant breakthroughs in human ingenuity are seen clearly by contemporaries, or achieved without controversy. It is instructive to go back in the history of computer science to consider two men who labored mightily pursuing ideas that once inspired nothing but neglect or antagonism in their lives.

The first of these was Charles Babbage, an engineer in

nineteenth-century England, whose life story is one unrelenting tale of rejection, ridicule, loss and frustration.

As a student, brighter than many of his instructors at Cambridge, Babbage put forward the notion that printed logarithmic tables, which were so difficult to compile and so shot through with mistakes, could be calculated by machine.

Not only could his proposed machine calculate the tables, it could print them out. (This was fifty years before typewriters.) In fact, this very remarkable machine he designed, called the Difference Engine, could "eat its own tail," generating results that would affect subsequent results being generated.

Babbage received a small grant to execute his design. Unfortunately, it was only about one-fiftieth the amount he needed. And money wasn't the only shortfall. Materials presented a problem. A machine as unprecedented as his required an estimated two tons of pewter, brass and steel parts whose custom design were beyond then-current machining capabilities.

In the midst of these setbacks, in 1827, when he was thirty-six, his father, wife and two of his children died and his own health began to deteriorate. To cap it off, his chief engineer quit, fired his crew, and took his drawings with him.

To put his prodigious mind to other matters, Babbage ran for a seat in Parliament. And lost.

At the end, Babbage would say he never had a happy day in his life. But his life was an unending stream of new ideas, one following on the other. Since the Difference Engine was now an impossible dream, there were no limits to how far he could scheme. And so he created the Analytical Engine.

Limited by the technology of his day, this machine was designed to rely on mechanical, not electronic, parts. But it was the prototype of every computer that has been built up to now.

The nomenclature evokes the Industrial Revolution. A "mill"

John Peers: At a Far Frontier

did the information processing, the number crunching. Meanwhile, the ''store'' served as the memory device, remembering a thousand fifty-digit numbers. The operating system was coded on punched cards whose openings were read by sensitive feeler wires. And, advanced even for today, the machine was designed to make conditional jumps in processing.

The Analytical Engine was never built, and Babbage died a broken, dispirited man—a visionary too far ahead of his time.

Two people did understand and back him in his lifetime. Ada Augusta, the Countess Lovelace, was a brilliant mathematician who saw the significance of Babbage's work and wrote on his behalf. Ironically, this first apologist for the computer age was the daughter of the Romantic poet Lord Byron.

Babbage's other supporter was the Duke of Wellington, who crushed Napoleon at Waterloo and would later, as Prime Minister, arrange for his governmment to give some financial backing to Babbage.

Seventy years after Babbage's death in 1871, people would finally begin to understand what he had done, and begin to build, from his designs, machines that were more than calculators, able to compute, to figure, to order, sort, analyze, reckon, process, recall.

Probably the first and certainly one of the most important people in this century to follow through on Babbage's ideas was a young professor of mathematics at Iowa State College (now University) in Ames. John Vincent Atanasoff was agonized, frustrated and tormented by the fact that the IBM tabulators of the late 1930s were too crude for his purposes. And when Atanasoff got tormented, he got *real* tormented. The story has it that he got so frustrated by his failure over several years to design an adequate calculating device that he decided to go out for a drink one night. And in "dry" Iowa then that meant a two-hundred-mile drive to Illinois. Atanasoff chose to pursue his destiny in a Ford V-8, at eighty miles an hour, one wild, winter night in 1937.

Once across the Illinois state line, he went to the first roadhouse he came to, ordered a couple of shots, and proceeded that night to design the blueprint for the electronic digital computer.

Atanasoff conceived of a machine that would use vacuum tubes and not electromechanical relays in its logic circuits, and hence be electronic; calculate not with base-ten numbers, but with base-two numbers—the binary one and zero; reckon in discrete units and so be digital; perform serial calculations and use capacitors for continuous regeneration of the electronic memory.

With a subsequent grant of $650 from the Iowa State Research Council in 1939, Atanasoff and an electrical engineering graduate student named Clifford Berry built the Atanasoff-Berry Computer. Unlike the room-filling ENIAC of a few years later, which would be put to productive use, the ABC was only a breadboard prototype, not much larger than a tabletop, and was never assigned to do a major project. But the sumbitch worked.

Atanasoff couldn't interest either of the major tabulating manufacturers then—IBM or Remington Rand—in his machine. For one thing, he refused to sign the standard release that all individual inventors face, which lets a corporation examine an invention with no commitment on its part, while forcing the inventor to sign away any future claims he might make against the company if it refuses his design but later produces something similar.

Years later when established computer companies were vying for positions in the field, they found it in their best interests to keep Atanasoff's name in the oblivion to which it had been relegated. And while Atanasoff never received

John Peers: At a Far Frontier

a penny for his seminal design, some corporate attorneys made their careers attempting to discredit his work. John Vincent Atanasoff was finally vindicated, though never financially rewarded, when a Federal District Court judge in Minneapolis, in 1973, ruled that Atanasoff was indeed the inventor of the first automatic, electronic, digital computer. Major corporate players in the industry brought out their best lawyers for the case. Atanasoff, for his side, merely produced in court every—*every*—document he had ever written. To the judge, anyway, the evidence was incontrovertible.

Why do I go back to the beginnings as I approach the end? Because the process goes on. The frustrating, painful breakthrough that was yesterday's cutting edge becomes today's convention. Babbage and Atanasoff were eventually vindicated, though for the former recognition came posthumously. There are others in our world today, living dramas that equal these. Their minds run too far ahead of us, or too much against the grain of today's conventional wisdom. Still they carry on, often alone. Some of them are, indeed, crazy. But some, in anonymity and painful silence, are obsessed with a solitary scheme which will be a *sine qua non* for the future.

I have noticed a few common traits about those whom I have met in the course of my work. Their glimpse of the future owes more to a hunch or gut instinct than to any academic credentials. Their training may have given them a better appreciation of what has gone before, and so they may be more rigorous in proceeding from here, but the goal, the scheme, the end, is generated more in the intuition than in the intellect.

For these visionaries success isn't assured at all, as it wasn't once for Noyce or Shugart. Their devices have no names now, as microprocessors and Shugart drives had no names once. Some will suffer that worst of all fates: to achieve the breakthrough, after years of work, only to come in a month behind someone else and see their independent brilliance relegated to footnotes in obscure technical journals.

It is easy to appreciate the insights of groundbreakers after the fact. It's something else to witness the self-absorption and the abruptness of people whose ability to formulate unprecedented syntheses and inferences has sometimes been developed at the expense of courtesy or coherence.

Yet however mind-boggling the realms these people inhabit—and however astonishing their brainchildren seem when they cease appearing in the pages of fantasy magazines and show up instead in patent application and trade journals—these people are "only" a part of a process, their work another ratcheting forward of the wheel of human ingenuity.

In the computer industry over the past forty years, for example, their work has meant progression from the accumulation of raw facts and numbers (data processing), to sorting and manipulating and analyzing those facts for business and social, as well as scientific, applications (information processing), to today's efforts to build machines that pull processed information together to draw inferences and extract syntheses ("fifth generation" knowledge processing).

One other point about the remarkable nature of these visionaries. They are, today anyway, as much businessmen as prophets. Many of them, here anyway, don't work under the protective wing of academia or the military or some supporting trust. Rather, while their work takes them beyond science fiction, they also have to contend with such matters as timely filing of W-2 forms, periodic reports to investors, selection of suitable vending services, the cost per square foot of office space, and the firing of a surly receptionist.

When I first met John Peers, he was a man in search of a company. He was incorporated, all right, and had a small staff and office, but he had no financing. His few colleagues talked in the other room when we met then, but it wasn't

John Peers: At a Far Frontier

clear to me whether they were engaged in productive research or idle time-killing until a grubstake came in the door.

A little over a year later, Peers and I met again. And again, it was a small crew and a small office. But this time there was a difference in the air: his Technology Industries had received substantial backing and was moving along with its plans to create the world's first recursive, or self-referential, semiconductor.

Briefly, that means that his product, should it prove to work, would be unlike any semiconductor ever, in its ability to say to itself regarding a given problem, "Well, I tried that and it didn't work. I think I'll come up with another approach." That requires special kinds of thought processes that have, so far, been reserved for humans.

The first semiconductor to begin this process was actually designed by one Charles Moore. Peers has put together the company to manufacture and market the product and license the technology.

When I asked several editors of trade magazines who cover the Bay Area what they knew of Peers, his name drew no glimmer of recognition, in spite of the fact that he has already started one Valley company, now ten years old.

Eighty percent of that company's business was done overseas, however, which might explain his relative obscurity here. In his native England, though, the *Financial Times* recently thought enough of him to devote half a page to his exploits in America.

John Peers was born in 1941 in the north of England. "Now I euphemistically refer to it as being of humble origin. At the time we thought of it as being poor." Financially unable to continue his education, he joined Procter & Gamble in the U.K. as a journeyman chemist at age fifteen. While subsequently employed in a number of jobs, he continued his studies and eventually took higher degrees in chemistry and physics.

In the late 1950s, he moved south to London and began to work in semiconductor production for the Philips organization. But soon after that, he decided he'd had enough of the sciences, and dropped out to become a professional musician, playing piano gigs in clubs.

The early '60s were a good time to get involved in music in England, but restless Peers decided after a year to move on again. He took a job with International Computers as a programmer for two years, and followed that with a stint at Singer selling business machines in London.

In the late 1960s, a Professor Morris Rubinoff of the Moore School in Philadelphia—the birthplace of ENIAC— backed Peers in starting a company of his own, "because he said that I should be an employer, not an employee."

That first Peers company marketed optical character recognition devices—a field "about which I knew very little." The company grew too fast for the capital it had and was sold to a Canadian group. A diehard entrepreneur, Peers then started another company in England called Allied Business Systems, which, within two years, grew to employ about 140 people. That year, 1973, would mark a turning point for Peers. The cutoff of OPEC oil, following the Yom Kippur War, had a devastating effect on England, as it did on most the industrial world.

"England is a country that is well known for pulling together in times of crisis. And sure enough, all the unions pulled together and went on strike at once. We ended up with no railway, no truckers, no dock workers, no electrical workers. I had a bunch of people manufacturing computers in the dark with no electricity and no customers. That's an interesting challenge. I joked that we tried to make hydraulic computers, but they all froze.

"That winter, I really had enough of England. Here was I, born entrepreneur, genuinely innovative in the field of computers, but England was not the place for me to blossom.

John Peers: At a Far Frontier

"England is a hotbed of invention, of capable people, of enormous enthusiasm. But it has a rigid structure that doesn't allow people who are 'lower-class' to be upwardly mobile. If you are born on the wrong side of the tracks and you don't have the right hyphenated name, it is tough to be successful. If in addition you come from the north of England, rather than the south, it's even worse.

"In America being an engineer is a highly regarded profession. In England a doctor and a lawyer are vastly different in social importance from a semiconductor engineer. The grubbiness of actually making something work is not socially as acceptable as using one's mind. I love England, but I believe that my attitude and hers are probably incompatible."

One British trait that Peers didn't relinquish is his "bloody persistence."

"I have the English bulldog insistence that once I believe in something you have to tear me loose from it. I will wear you down if I don't intellectually convert you."

In 1974, he left the U.K. and relocated in Silicon Valley. Having invented a multiterminal business system, an operating system and a type of a computer terminal in his native country, he set out here to invent a new kind of business computer that could be programmed by the users and didn't require professional programmers. After all, he figured, the user knew his requirements best and should be the one to make the computer adapt to the intended use. It didn't make sense to bring a professional programmer into a company to tailor the business to the machine.

Peers's new computer was, according to him, the first computer in the world that didn't require that the user know English. True to the notion of user control, if the user first started his dialogue with the machine in, say, Portuguese, the machine was Portuguese from then on.

Peers was chief executive of his company from 1974 until

the end of 1979. He had drawn up a plan to make a significant acquisition to expand the business which, he says, would have been enormously profitable, as subsequent events bore out. Alas for Peers, he ran afoul of the company's financial backers.

"They decided I was a spendthrift and a rotten manager and I was running away with their money. I understand their point of view, but unfortunately their running for the hills caused them to lose a great deal of money."

In late 1979, Peers was fired as a result of the disagreements, chucked out of the company he had formed.

"I had gone through a divorce in 1978 and was under quite severe pressure. A divorce in the Western United States is financially a very debilitating event. Thank God, I had a reasonable wife who asked for what was perfectly reasonable at the time, although neither she nor I foresaw the events of 1979. She got the real property, and I got participation in the company I had formed, which seemed a fair split at the time. Unfortunately, it turned out that the equity in the company wasn't worth very much, because these people were trying to take it away from me.

"They controlled the company, and they were suing the company which they controlled. Every time we asked the company to fight back, they had a shareholders' meeting and said the company shouldn't fight back."

He was soon out, and "between engagements."

Peers's fascination with a self-referential computer, one that is aware of its own operation in order to adjust that operation on its own, led him to an involvement with Forth, Inc., an already-existing California-based company which created a recursive software language called Forth. His Technology Industries would grow out of this involvement, the intention being to produce both recursive software (Forth) and hardware (the Moore semiconductor).

*   *   *

John Peers: At a Far Frontier

Before going any further, a brief digression is necessary on the subject of artificial intelligence, or AI, since it will come up from now on.

AI represents the next generation of machine intelligence, usually referring to software but in Peers's scheme available in the hardware too, that can draw inferences and conclusions, appears (by virtue of its processing speed) to have intuitive capabilities, can recognize relationships, synthesize several concepts to form a new idea, respond to unique situations, and maybe someday laugh at—or even tell—a joke. In short, a machine running under a sophisticated AI program would appear to think in a way we would currently consider human. This "heuristic" intelligence is opposed to the algorithmic processes of traditional computers, which deal with information in a very rigid, lockstep sequence that allows no deviation from the program.

HAL, the famous fictional computer in *2001: A Space Odyssey*, drew his acronym/name from his Heuristic-ALgorithmic dual capability. He could step through the algorithms that controlled the proper functioning of the life-support systems on the spacecraft. He could also make a decision on his own to terminate those functions when he thought the astronauts were threatening him.

The field of AI has developed at three centers in this country: under the direction of Marvin Minsky at MIT; under Allen Newell and Herbert Simon at Carnegie-Mellon University; and around John McCarthy and Edward Feigenbaum at Stanford. The term was coined by McCarthy back in 1956, though the subject has only recently begun to receive much widespread coverage.

The spiritual godfather of the field was an extraordinarily gifted young English mathematician and logician named Alan Turing (1912–1954) who, in the mid-1940s, first posed the question of whether a machine can actually think. Heuristic computers, in fact, force us to think even more creatively, to wonder, if a machine can think like us, is its intelligence really "artificial"?

While digressing, and because this subject will also come up again, we should consider, in absurdly brief fashion, the field of genetic engineering.

In 1953, two Englishmen, James Watson and Francis Crick, uncovered the double-helix design of deoxyribonucleic acid (DNA), the coding mechanism that carries hereditary instructions from one generation of organic life to the next.

A strand of DNA, called a chromosome, is made up of millions of genes, and each of these genes contains about 1,500 so-called complimentary base pairs.

Jumping ahead to the mid-1970s, molecular biologists discovered they could take a gene out of a DNA strand and replace it with another gene. This genetic recombining, or recombinant, process would, for starters, allow the natural "manufacture" of quantities of such needed, but rare-in-nature, drugs as insulin, human growth hormone and interferon, often using far less energy and producing less waste than are inherent in the synthetic chemical production of such drugs. The ultimate uses of recombinant DNA technology are no more possible to project now than are the ultimate uses of computers.

In the early 1970s, a professor of molecular biology at UC San Francisco named Herbert Boyer, working with Stanley Cohen of Stanford, got the patent on the basic recombinant process. Soon after, Boyer and financier Robert Swanson formed a company in South San Francisco called Genentech, which followed the opening of another company that would also pursue similar work near Berkeley called Cetus. An industry was born.

Returning now to John Peers, it was difficult for him to get financing to start his Technology Industries, to design an

John Peers: At a Far Frontier

inferential capability in silicon and make an intelligent semiconductor. For two reasons: first, there was a question mark after his name from his former business; second, venture capitalists generally prefer to back companies with products ready to bring to market, rather than support basic research.

"In 1982, I determined I was going to have trouble throughout my life getting people to understand the way my mind works, and getting them to understand I'm not a madman out of control. I may have a perspective ahead of my time, but I really do think things through very carefully. And what looks like hip-shooting is not hip-shooting. For some reason, the majority of people with whom I deal do not perceive the amount of sweat and toil that goes into what I do.

"The majority of venture capitalists think that I am a spendthrift, a rotten manager and a madman, and would not give me a penny. There are exceptions, but that is the majority. And the majority of that image comes, unfortunately, from the word put out by the people in my former company who went after my reputation, as well as my soul and various parts of my anatomy.

"It was a tough five years. I didn't get married to get divorced. When you get divorced, when you've lost your house and your kids and your wife and your company . . . now somebody wants you to lose your self-respect, your honesty, your integrity . . . what do you have left? I was fortunate to have some friends who gave me the belief in myself to reinforce this little, funny core that sits in the middle of me which says, 'No, they are wrong. And I am *right*!' "

Finally, a man name Salam Qureishi, who has a successful Cupertino-based company called Sysorex which sells computer systems management to the U.S. and foreign governments, invested a substantial amount of money in Peers's project. (It was decided that in order to focus the business, Technology Industries would divest itself of Forth, Inc.)

According to Peers, if the semiconductor works to spec, it will offer one-fifth the computational power of a Cray supercomputer for 1/2,500th the cost. Let's be more precise: a computer made up of a single board measuring six by eight inches, containing this chip and supporting circuits, would sell for $4,000 and possess 20 percent of the power of a machine that now sells for over $10 million and is the size of a small armoire. Its graphics capability, for example, would allow the user at home to create animated images similar to a Walt Disney–style production in their color, depth and perspective.

The computer would be extremely fast, extraordinarily easy to program, able to change its functionality as it goes along and capable of speech recognition. More significantly, it could be "chained," that is, linked to similar machines, to build a mainframe computer out of several microcomputers. That is not how it is done today; a mainframe computer is not a lot of little computers hooked together.

Peers makes one last point: the chip could emulate a neuron. That is, if you put enough of them together, until you have a critical mass, you might have, in other words, the beginning of a neurological array. You might have, in other words, the beginning of a system that learns on its own!

Oh, and there's one other possibility. The chip may not work at all.

"Today we make computers out of precise elements, and the system ends up not just precise, but rigid. We humans are precise but flexible. If I am right about the use of this chip and its progeny, we will end up creating a system that is precise and flexible, made out of imprecise units."

Peers then tells me about some current thinking regarding how the mind retains information. Clearly we can't contain a binary representation of everything we've experienced, yet we can recall so much with such clarity. There must be a massive data-reduction scheme at work in our brains that lets

John Peers: At a Far Frontier

us carry so much information, so many vivid memories, and pack it all under a size seven fedora.

What if we could emulate that data reduction in machines that work not at our millisecond speed, but at machine speed, nanosecond speed? We then, Peers says, have real augmentation to ourselves. "By amplifying our own self in the way we are made, we become more than we were."

And is there then, I wonder, some way to merge this with recombinant DNA technology?

"Perfectly valid question. You will find in the latter part of this century all the recombinant DNA work, all the microcoding, all the brain decoding, all the social sciences' group behavior theory, all the awareness sciences, all the Eastern philosophy, all the theories of learning, all the theories of communicating . . . all coming together into one new science we can't name yet.

"The computer revolution has got nothing to do with creating machines that do invoices. It has got everything to do with human destiny. The amplification of who we are and the joining of us with what has to come next. A oneness.

"We are a function of what we perceive, and we are about to change the tools by which we perceive. And that joining of us with the new tools will give us a new perception as to what tools we will then have to make to change our perception. It is a never-ending course. A never-ending expanding one with the universe."

To Peers, the history of the universe is a story of aggregation. In emptiness there occurred a Big Bang. Lots of energy was present in the form of waves, which clustered into areas, which became particles, which agglomerated into protoatoms, and on it went . . . many things coming together to form a new entity which is more than the sum of its parts. And now . . .

"We are almost at the point where we've got enough human beings and enough knowledge to cluster into something new," as yet unknown and unnamed.

Among the possibilities: the ability for the next generation to live an average of 250 years, as our technology lets us duplicate and replicate those organic parts which wear out, up to, and including, the brain.

Peers gets incensed when I use the word "synthetic" to refer to these replacement parts. After all, he says, there isn't an atom in your body that was there two years ago, or that will be there two years from now. Yet you're still the same "you." "You are who you are, but you are not made up of what you were made up of."

I can accept the idea of brain transplants. I have trouble accepting the ramifications of it.

"If you have extension of your own memory, because your memory is beginning to fail, and you fill that 'synthetic' memory—to use your words—with your stored recollections, that memory is an indelible part of who you are. You have externalized yourself by becoming symbiotic with something. Right?

"If you have your memory externalized because you have a replacement part that isn't you, that has taken over from your failing organic part, you have clearly externalized your personality. And that system can't have a limit, because it can be connected to other systems, to give you access to all sorts of other information.

"But how are you to know whether those other systems that your system is connected to aren't also connected to another bank of memories that happens to be another human being?"

Share my memories with strangers? Risk blurring the finely honed edges of this ego I have taken a lifetime learning to protect? There have to be limits to what we can attempt. Limits to what we can know.

"There is no limit. The marvelous work of Gödel says that you cannot define a limit, because the minute you put a limit in place you can invent beyond the limit. No matter how

John Peers: At a Far Frontier

all-encompassing the limit is, it defines the point at which you can exceed it. That's all it does.

"If you look at the history of man, every time we got to the end of a century, we said, 'We know ninety percent of everything there is to know.' And every hundred years we look back and laugh and say, 'They didn't know, but we do.' And we've done it every century now! And we still haven't learned! We're still here today saying, 'Now we know ninety percent of everything. We're down to quarks. We're down to mesons.' And it's no more the end of the road than it was a million years ago."

"We live in a society dominated by violence . . . in the press, on television. But the greatest joys that I know in my life have got nothing to do with violence. They have to do with skill and knowledge, with the love of another human being. It has to do with reading a marvelous book, listening to a perfect piece of music, enjoying a fabulous meal. From understanding something. There's a leap of consciousness, and a new piece of you comes into being. The more we know, the more we are. And the more we are, the more we want to know."

This is all well and good, but we have moved from business plan to blue sky. How does Peers propose to use his semiconductor to support his philosophy?

Initially, he says, the device will be used in a conventional manner. It will be very powerful and low-priced, but will require less circuitry and less electrical power than standard chips. In addition, it will be relatively easy to program, doing things that can't be done simply today.

After that, according to Peers, people will begin to see the real breakthrough this chip represents. "Because it's adaptive. The minute you make something that can change its discrete functioning at the time at which it is performing what it's performing, then you can make a circuit that can refer to itself. You can make a circuit that can perform a function, test whether it did it very well, change itself and do it again.''

By conventional standards, this semiconductor is very low-density, containing only the equivalent of about 15,000 transistors. By comparison, some integrated circuits today carry the processing power of a half-million transistors.

The difference between this semiconductor and a traditional integrated circuit is like the difference between two kinds of transportation.

''Every semiconductor made today has its function designed before it is ever made. It's like a train. You're making an engine, and you've got to define the width of the wheels, the weight it can take, the speed it can travel, the number of carriages it can pull. And you'd better define those pretty exactly before you make it, because if you get it wrong it ain't gonna fit on the track and nobody's going to use it.

''I believe we've made a motor car. Where, instead of defining the course for it, we put it on the road and we teach people to drive. After that, what it does, what it pulls, where it goes is no longer a question for the designer. It's a question for the user.

''In modern-day computers, there is programming, and there is the data the computer runs. The organic system has no differentiation. The data implies the programming, the programming implies the data. They are one and the same. It is the use to which they are put which describes their function at the time. We have artificially separated those things in computing so far. I believe we are on the verge of remerging them.''

That ''artificial separation'' is called Von Neumann architecture, after a brilliant Hungarian-born mathematician named John von Neumann who, among many other things, collaborated with J. Presper Eckert and John Mauchly on ENIAC's son EDVAC. Whereas ENIAC's operations were controlled by patch cords that had to be reconfigured for each program,

John Peers: At a Far Frontier

EDVAC's programming was stored electronically, similar to, but distinct from, the data being processed. This critical separation of program from data, of process from content, has survived to the present as *the* architecture of computers.

Peers thinks that there is an exit out of what he sees as the cul-de-sac of Von Neumann's scheme.

"You cannot have buses and you can't have addresses in a computer that would think the way we do, because that isn't how *we* work. A holographic memory, such as humans have, does not work like a computer memory. If you cut out a piece of holographic memory, you don't lose the image, you only lose a bit of exactitude. You still keep the sense of what the picture is. But if you lose a couple of bits out of a computer program, you run riot.

"My point is, the new machines—of which I believe this chip is one—will begin to blur the distinction between program and data. And if we're right at all, we're right totally. If this chip works at all, it is only a question of time before it works exactly like we humans work. And if I don't do this, somebody else will. I'm not blowing my own trumpet here. I am saying the way we have gone so far will not lead us to where we have to go. We have to have a new method.

"We have to have machines that, as a side effect of functioning, change their programming. Because they don't have any programming to begin with. They only have a data base, and the data base relates to itself. In exactly the same way that we base our knowledge on twenty-six letters that have no meaning on their own and represent both program and content. Once we understand the creation of a structure which has meaning because of the way we make it, then we'll understand how to proceed to the next level of meaning. And we will then be symbiotic with that machine."

All well and good, but I still want to get down to practical applications.

"Right now, if you walk down any street you find guys with their Sony Walkman headphones on. It won't be long before we have a computer we talk to, and that talks back to us. What we're probably going to do after that then is have a transducive pickup which is attached directly to the bone of the inner ear, so that instead of talking out loud to us, the computer just talks back to us by some transducive form going straight into our nervous system."

Where does that take us?

"There are things you can't do with words. You can't describe paintings with words, you can't see movies with words. And so you are also going to be able to tie this computer into the optic nerve, for example. At that point, it begins to blur as to who you are and who you're not."

Do you mean, I asked Peers, that if I wanted to see *Casablanca* on a moment's notice, all I'd have to do is close my eyes, and there it would be, willed up from some massive data base that's always at my disposal?

"You wouldn't have to close your eyes. The machine is an extension of you. If it is constructed like you are, where does 'you' end? I have capped teeth, but I don't think of them as capped teeth. They are mine. When I chew something, the ends of the teeth are mine.

"The perfect tool is transparent to its user. The telephone handset is a good example. When you are using it, you're not aware it's there. You wait until we get a biofeedback circuit that is a piece of intelligence which is us, but is not biologically a part of our evolution. When you lose that you're going to feel really rejected. Because your new normality will include this power that you now have. When you go back to being what you were, you will feel lessened."

Even as he describes a rather monumental branching on the evolutionary tree, all based on his product, he retains a typically droll sense of humor.

"Remember all the comments I'm making today are highly biased, and could be totally wrong. But that's the joy of living. . . ."

John Peers: At a Far Frontier

That momentary display of whimsy having passed, he proceeds.

"Once you have externalized part of your memory, you can now think into a system which contains information which you have not yet learned.

"Right now, you can sit at a computer terminal and ask for data and it gives it back to you. You don't know the answer until you get it. Translate that into a memory system which is endemic to you. You think into it and it gives you back information which is now part of you."

And what memory—or data base, or warehouse of information—you can't carry with you, even in some greatly reduced form, could be supplemented by a larger data base that is remote but accessible, since you would have your very own sending and receiving device always on you, and your own radio frequency.

"You'd think nothing of having the entirety of the *Encyclopedia Britannica* in your own personal possession. Your perception of normality will change."

I daresay.

"You'd think into the system and it thinks into you and says, 'Do you want supper at six?' And you think back into it and say, 'Yes, I do,' or 'No, I don't.' And it says to you, 'What do you want?' And you think into it, 'I'll have hamburgers tonight.'

"Or this. The system says to you, 'I've been thinking about the work you were doing yesterday on particle physics and I think you're on to something.' And you think back and say, 'Yeah, but I'd like to know more about suchandsuch.' And there it is for you."

But how do you know whether that question is coming from the endemic system which is part of you, or from another system under control of another human being? What happens to the notion of self that we have spent thousands of years developing if (or when) we reach a point where we don't know if we are thinking our own thoughts, or a machine is thinking them, or another source—human or machine—is thinking them?

"It won't matter whether you're worried about personal identity or group identity, because it will be one and the same. In exactly the same way as you can't say which nerve cell is the one that causes you to be who you are. In an extended sense there will be no difference.

"It won't matter which nerve cell causes you to be who you are because you will be bigger than you were before. And it doesn't matter if part of you is shared with somebody else, as long as you have access to it when you want it. Because that is you."

But what's so wrong with the ego as it is, that we want or need to replace it?

"One of the greatest challenges of who we are is loneliness. And a great deal of ego is the result of having to put the ego up to balance loneliness. If we no longer have loneliness, I'm not sure ego will still be that necessary.

"Let me try to clarify it a little bit more. One of the beliefs I have is that once we automate the majority of functions we spend the bulk of our daily lives doing, we will have more time left on personal development. One of the things I'm forecasting, which may seem strange at first, is that we're going to end up spending a lot more time dealing with emotions, psychological phenomena, as a result of being more mechanized. It's because we're mechanized, we won't spend the time being mechanized. And if, as a result of that, we have more time to deal with emotions, surely we'll have more ability to understand who we are, and deal with these rushes of emotions which are currently beyond our control.

"The more you understand yourself, the less your ego gets in the way of experiencing reality. That's my observation about life. The more insecure people are, the more they have an ego problem."

Insecure or not, I find it hard to abandon ego. Ego, after

John Peers: At a Far Frontier

all, is the one thing that will never abandon you. Health, money, friends can all depart you, but you can always fall back on yourself. That's a central theme in Western culture, from Ulysses to Sam Spade. To be told that the final source of security is insecurity . . . that's unnerving and humbling. And there's more humbling to come.

"My experience of intelligent people is that the more they know, the more there is an increasing awareness of how much they don't know. The more you know, the humbler you get. Because your perception of what still remains to be learned increases faster than what you learn. With these new tools, we are going to increase the rate at which we are humble."

But why should knowledge be humbling? Peers's reply is to tell me that I have stumbled over Peers's Law. That is that the solution to a problem changes the nature of the problem. "It will go down in history one of these days."

Peers's Law, he tells me, is the macro world equivalent of the Heisenberg Uncertainty Principle, which holds that, regarding the electron, it's hard to measure the thing without changing what's being measured. Peers would take that a step further, and say that not only is the object of our investigation changed, but the measuring tool is changed as well: as we learn something about the object, so are we changed as a result of that new information; the newly informed us, therefore, now needs to know even more in view of our new state. It's a never-ending cycle: the solution to a problem changes the nature of the problem.

Isn't that a bleak vision? I ask. There is no end of problems then. We no more than solve one and we create a new one that demands attention.

Peers sees this in a more optimistic light: the glass isn't half empty, it's half full.

Peers says he is not judging whether a picture such as he paints would be right or wrong; he is, however, sure it's coming. And its coming, he says, revealing which side he takes, is "joyously inevitable." We use our minds to create tools, with which to bootstrap our minds to new levels, from which we take sightings of new horizons for our minds to aim at. He calls this "externalizing evolution."

"We cannot organically change fast enough to keep up with the demands we're placing upon ourselves. If you look at the rate of evolution, it's shaped like a hockey stick. We can't organically change ourselves fast enough, but we're inventing problems that need ourselves to be changed."

And to leave the change to biological evolution is no longer possible, because that proceeds at a glacial pace compared to the speed with which we need to respond to the exploding amounts of information which technology has provided for us.

Our technology gives us new information; we need a new technology to augment our native ability to deal with what more we know. The solution to a problem changes the nature of the problem. Hence, for Peers, the inherent limitation in traditional computers which need to be programmed with information that contains the answers if they are to give us the answers to questions we ask of them. Hence, further, the need for self-educating, inferential computers that have enhanced speed, as well as ability, to help us think in new directions, in order to help us keep up with the new directions we keep discovering.

"The majority of people regard me as a maniac. If not a maniac, certainly dwelling too far in the future." And, with that, Peers walks me to the door and I step out into the dark in his now-deserted office park.

It's becoming clear that Silicon Valley is really engaged in something more than electron manipulating and deal making. At its center, it isn't simply engaged in designing and marketing widgets. The real obsession here, whether it's recognized

John Peers: At a Far Frontier

or not, is identifying in a computer—in a machine—the quality that most makes us us: intelligence.

The underlying fascination of Silicon Valley—both within and without the Valley—is the process of putting into, and extracting from, machines the attributes that, until well into this century, we thought of as unique to humans: analyzing, computing, comparing.

Parallel with that, it turns out—not coincidentally—work also goes on here searching for signs of intelligence in another unexpected place: among the stars.

John Billingham: Searching for Intelligence Amid the Stars

Someone new to this area could drive the Bayshore Freeway from Palo Alto to San Jose and never be aware that he or she had just driven through the Silicon Valley. There aren't a lot of distinguishing signs. In fact, the most extraordinary sight along that stretch of U.S. 101 has nothing to do with Silicon Valley at all. It's the astonishing sweep of open field and gigantic blimp hangars that make up Moffett Field Naval Air Station in Mountain View.

Yet adjacent to Moffett and only slightly visible from the highway, at the National Aeronautics and Space Administration's 360-acre Ames Research Center, scientific and technical work goes on that is very much in the exploratory spirit of Silicon Valley: a search for the nature and location of nonhuman intelligent life.

Behind the secured fences at Ames, men and women carry out research in a number of fields such as astronautics and space medicine. These people, according to their job classifi-

Searching for Intelligence Amid the Stars

cations, are civil servants; they move through government corridors, fill out government forms, no doubt deal with a certain level of government inertia, and eat lunch outside on government picnic tables amid wind tunnels and flight-simulation complexes. Yet one small program there could, conceivably, be of more far-reaching impact that anything else being carried out on earth.

For among the research work conducted at Ames is a search for extraterrestrial intelligent life, a quest which goes by the acronym SETI.

The chief of the Extraterrestrial Research Division is a charming, mild mannered Englishman with a background in aviation medicine who joined NASA in 1963 and was subsequently involved with the Mercury, Gemini and Apollo programs in Houston before coming out here.

Dr. John Billingham speaks of his work with the calm precision one might use to describe the route to Sacramento. Maybe that's what one has to do to deal with issues like these on a daily basis. But there's no mistake about it, under the placid demeanor he is keenly aware of the potential importance of his work.

Once, in our conversation, he used the term "galactic," and it hit me like a fist. The word has been so overused in science fiction that it's lost its impact. But here is a man using it for its precise, scientific meaning. Here is a man who really deals in galactic affairs. And drives to work on the Bayshore Freeway.

Before he began to tell me about SETI and how that connects with the work of Silicon Valley, Dr. Billingham thought it would be worthwhile to give me a brief overview of NASA's current comprehensive study of the origin, evolution, nature and distribution of life in the universe—a field called exobiology.

*   *   *

Searching for Intelligence Amid the Stars

In the beginning was the Big Bang, some fifteen billion years ago, the primordial explosion from which came into existence about 10^{10} galaxies, about 10^{22} stars within those galaxies, and who knows how many planets around those stars.

Eventually, basic atoms became more complex atoms, and certain of those atoms, such as oxygen, hydrogen and carbon, came together to form organic compounds. This may have happened on quintillions of planets. Or maybe not. All we know for sure is that it happened here on earth. And these organic compounds assembled themselves into more complex arrangements. And they began to reproduce. And by and by, these organic compounds became what we would recognize as living systems. That happened about 3.7 billion years ago on this planet.

For the next 2.5 billion years, life on earth consisted solely of these single-cell creatures, although as time passed their internal biochemistry evolved and became more sophisticated.

Then about a billion years ago, for reasons still not clearly understood, though perhaps because of the genetic diversity that came with sexual reproduction, there occurred an explosion in the evolutionary process on earth. Single-celled organisms started to become multicellular, and diversified into fifty or more major families. Eventually some fish became amphibious, and then there were reptiles, mammals, primates, man. As living creatures had become more complex, so too had brain processes, until, in man, there appeared an animal that could conceive in abstract terms, could make models based on those conceptions, and could communicate the conceptions and the models to others in the tribe.

From this point on, the study of the development of intelligent life on earth shifts from biological to cultural evolution. Whereas before, a species had to wait tens or hundreds of thousands of years for changes to occur because of mutations, now there were young individuals with flexible, receptive brains that were waiting to absorb models passed on by their

Searching for Intelligence Amid the Stars

parents and teachers. Models, Billingham says, that let them predict to some small degree the future.

"Let me explain what I mean by predict the future. If you have a little bit more brain matter than the tribe over the hill, you may be able to figure out that when the wind blows from the west at certain times of the year, you will find much game at the waterhole which is twenty-five miles away. And so it's worth your while to make the journey to that waterhole, because you know you'll find a rich harvest of food.

"If you didn't have that little bit of extra gray matter, to allow you to build that predictive model, you waste your time going back and forth. Or you don't bother to go enough times, so you don't survive. What has survived is the most successful of all the experiments."

In the last two million years, our brains have increased threefold in size. And with humankind's ability to model the external world becoming so well developed, toolmaking and technology were introduced. This further advanced cultural evolution even more as men and women learned to husband resources, including their own wits, to stay warm, well fed, healthy. Then came the development of spoken language, symbolic communications, and an even more sophisticated refinement of that, written communications. And so we come to the beginning of historical time, about ten thousand years ago.

A major question which Billingham addresses in his work is this: Could the chain of events that occurred on this planet have likewise taken place elsewhere in the universe?

And given that there are 10,000,000,000,000,000,000,000 stars in the universe, many of them possibly possessing a retinue of planets, life need exist on only a tiny, tiny fraction of those planets for the total number of life sites to be in the billions.

"We think that life is widespread in the universe, and intelligent life is also widespread. And if that is the case, then you have to ask the question, is there any way of detecting this life?''

Currently, two methods are being employed to answer that question. The first approach is to look for life in our solar system, which the Viking project set out to do by landing two spacecraft on Mars in 1976 to detect microorganisms. None was found.

The other method is to search for life beyond the solar system. And this search must necessarily be for intelligent life, since the distances to the other stars are far too great for us to be able to send interstellar Viking missions at this time. The search for intelligent extraterrestrial life centers on the search for the existence of radio waves emitted from other civilizations.

The first part of the SETI program is to research, develop and build a prototype of the world's best spectrum analyzer and signal processor, to split up the incoming signals into millions of separate channels, in order to spot a signature of intelligence. That is currently being done now at the electrical engineering department at Stanford by Allen Peterson and Ivan Linscott. The second phase, which will begin in a few years, will be to put that analyzer/processor into operation at large radio telescopes, scanning the heavens in search of signals in the microwave region of the electromagnetic spectrum between one and ten gigahertz.

It strikes me as a sort of coming full circle. The EE department at Stanford, which grew out of Fred Terman's fascination with radio communications, is designing a tool to listen for the music of the spheres.

Too, the man recently brought in specifically to head the SETI program, within the Extraterrestrial Research Division, is Dr. Bernard Oliver, formerly vice-president of research and development at Hewlett-Packard. With Billingham, Oliver cocreated the program some years back, and had served as a consultant to it while at H-P.

* * *

Searching for Intelligence Amid the Stars

If, after all the years of effort, a signal from an extraterrestrial source should be detected, what would Billingham's reaction be?

"Disbelief."

That signal, maybe a thousand years in transit—should it ever be detected—would undergo the most scrupulous testing possible, from the most advanced analytical devices we have. The first tests would determine if the signal was simply radio frequency interference (RFI) from earth, a glitch in the software, a spurious, wayward signal, or a hoax.

If the signal detected by one receiver passes all these tests, then another observatory in another part of this suddenly smaller world would be asked to train its radio telescope on the given coordinates, at the given frequency, with the right sensitivity.

If that instrument, too, detects the signal, then radio telescopes around the earth would lock on the coordinates for further verification. The dry, monotone conversations of scientists at outposts around the planet, speaking in their opaque jargon, could not mask the breathtaking import of the discovery: we are not alone in the universe.

The signal could be either from a "beacon," intentionally set up to send a content-rich message informing anyone listening in of a civilization's existence, or it could be a signal that inadvertently escaped a planet, a radio message intended only for reception on that planet or in that solar system. In the same way that for the past sixty years or so, since the beginning of large-scale radio transmission on earth, we have unintentionally been beaming the fact of our existence out into the universe. Our earliest radio programs are now sixty light-years from earth, bound for infinity.

"It is possible that if there were another civilization out there, they might already have heard us. And they might have replied, although it's not terribly likely. In which case, if they did, there might be a message coming back to us. But it would only be from the nearest stars, because the ones farther out would not have heard from us yet. So we will look at the nearest stars first.''

But isn't this, I wondered, the ultimate search for a needle in a haystack? However exciting the prospect, however probable the existence of a beacon out there, how likely is it that one of those signals will pass our way, at the same time that one of our listening devices is tuned to that frequency and those coordinates?

With marvelous British aplomb, Billingham replies, ''Well, one has a certain hope that one may be lucky. That is what it boils down to.''

Actually, the SETI program isn't quite as quixotic as it might at first appear. The NASA spectrum analyzer/signal processor prototype, currently being built jointly by Stanford and the Jet Propulsion Lab in Pasadena, will be able to analyze 74,000 frequencies simultaneously. And that is only to test the design of the ultimate device. In the early 1990s, when SETI moves from R&D into full operation, it will use a device that is scanning and analyzing a hundred million channels at once.

''The rate at which progress continues to be made in electronics and very-large-scale integrated circuitry, and with the variety of different new techniques for signal processing . . . a lot of the work being done here in the valley . . . all this holds out considerable hope that the odds against us are rapidly going to drop. The odds in favor of success are going to rise significantly within my lifetime.''

The signals may always have been there; it may only have required the development of sophisticated listening devices to detect them. That reality may indeed come to pass in the not-too-distant future, although funding for the project is rather small (currently only about $1.5 million a year).

Money, alas, and the laws of physics are universal barriers, which is why Billingham doubts we've ever been visited by

Searching for Intelligence Amid the Stars

UFOs. The costs of sending a machine across the enormous distances of space are—is it surprising?—astronomical. A trip to the star nearest to us, only four light-years away, would take 500,000 years' worth of the total U.S. power consumption.

I think that of all the questions I asked in the course of this project, no answer surprised me more than Billingham's response to my query: If you detect a signal, how would you reply?

A most emphatic: No reply! This is a "search for," not a "communications with" extraterrestrial intelligence, he tells me.

There are some people who are very much opposed to this whole program, on grounds that to reply is to give away our position. For two things are certain about intelligent extraterrestrials who have been beaming signals for some time: they don't speak English, and they are much older, and much more advanced, than we are.

If they have, say, a hundred-thousand-year jump on us, that makes them the Europeans and us the natives. By several orders of magnitude. Fine for them; maybe not so for us, in some people's opinions.

On the other hand. Since the discovery of radio technology and nuclear fission occur at about the same time (at least they did on earth), any civilization that has been beaming signals for, again say, a hundred thousand years before they sent one our way may have an equal amount of experience in living with that and other threats to their survival. And they may be able to teach us quite a lot.

Another possibility is that they may have become extinct, having destroyed themselves in the years the signal was in transit, because they couldn't control the forces they discovered or for some other reason. We might listen in and learn a lot from that, too.

Whatever the risk or payoff, NASA's charter is limited. The question of presuming to reply for all humanity is too big a decision for one agency of one government to make. "If the signal were to contain lots of nasty threats, maybe nobody would want to reply. But if the signal were benign and informative, which I think will be the case, perhaps a reply, then a dialogue, would be of great interest and value. But that decision would have to be taken, in some way, by the people of the planet earth."

The whole notion of communicating with extraterrestrials is the story of the ultimate culture shock. Yet consider how, over the last thirty years, pop culture has come to accept the idea. From the gnarly antennae and malicious habits of "aliens" in the '50s, to the impersonal intelligence represented by the black monolith of Stanley Kubrick's *2001*, to Steven Spielberg's lovable *E.T.*

Having devoted all these years to the subject, what does Billingham hope will come of it? What does he think the messages might contain?

"All questions about extraterrestrial life are open. One wants to be hopeful and say, yes, we'll detect them and, yes, they are sufficiently advanced that they have long since got past the sort of primitive stage we are at. And, yes, they have dealt with the very problems that we face here, without doing away with their spirit and individuality. And, yes, they are able to tell us their own history.

"If that were the case, we would know for the first time if somebody has been this way before and has achieved a stability over a very long period of time as a civilization. They might be able to tell us about some places where they went wrong and very nearly blew it. Or they might be able to tell us about the history of other civilizations which were part of the galactic community—or whatever you want to call it—which got to our stage and then did something which was catastrophic, fatal. In other words, communicate to us what

Searching for Intelligence Amid the Stars

are the ingredients of success over a very long period of time.''

There may, in fact, be a network of advanced civilizations, thousands, millions, perhaps billions of them, communicating with each other across the universe. And us perhaps soon to join the network, the new kid in school, our most sophisticated technology as wooden clubs to that crowd. Wait until the phone phreaks hear about this network!

Billingham acknowledges that communicating over such great distances takes time, perhaps several generations. But even if the optimum situation doesn't unfold, we will still have learned quite a lot from the experiments. Various NASA missions have shown that organic chemical processes go on out in space. By extension, then, organic chemistry may be a universal phenomenon. And even if the message we ultimately receive is from a now-dead planet, the one-way communication would be important. ''We communicate all the time with the ancient Greeks, and spend a lot of money on it. And they weren't even an advanced civilization.''

After Billingham explained SETI to me, we returned to the subject of evolution again: Where do we go from here?

Consider the breakthroughs we have had in this century, in radio communications, air travel, machine intelligence and nuclear fission. These are small change, he feels, compared to what he sees as an upcoming two-pronged revolution of ''the same magnitude as the revolution caused by the change from simple genetic evolution to cultural evolution.'' That is the change that will start with the fuller implementation of artificial intelligence.

''This will impact our own evolution as human beings because it makes available vast increases in knowledge, without necessarily any change in our anatomy, physiology, appearance or brain size. And there are no fundamental barriers to artificial intelligence. It can become at least as capable, if not more capable, than our own intelligence. That's the one breakthrough which I'm sure about. There's another one which I'm not sure about.''

Billingham perceives that second breakthrough, in genetic engineering, as a potential threat, but it's not clear yet if the threat is to our future or to our present sense of who and what we are.

''It's conceivable that someone, if society would allow it, could alter our own genetic instructions so that one might, perhaps, increase our innate level of intelligence. Now that raises the specter of all sorts of science fiction types of evolutionary changes. We would be deliberately interfering with human evolution, and not allowing it to take a natural course. And you know that raises screams of anguish almost universally. Using recombinant DNA techniques to create a larger skull, with correspondingly more brain matter, for example.

''On the other hand, that may not be possible to do. And I am certainly not advocating it! You see, we don't even know the limits to those sorts of things.''

There is nothing immediately apparent in Silicon Valley semiconductors that makes one raise fundamental issues, questions of first causes and final ends. But semiconductors *are* part of the pattern of our evolution. From NASA's study of a void, of a Bang, of carbon, hydrogen and oxygen atoms being formed and later forming protoplasm, to creatures that exist, that move, that think, that communicate. . . . From the rudimentary experiments in wave generation by Hertz, to Shockely's transistor, to Noyce's integrated circuit, to Hoff's microprocessor, to Peers's semiconductor (if it works). . . . From Terman's radio lab to Billingham's search (if it succeeds). . . . From Babbage's early-Victorian computer, to Atanasoff's breadboard mockup, to Turing's prescient question, to Forth language. . . . From Mendel's experiments with peas, to Watson's and Crick's discovery of the double helix, to industrial-scale production of the fluids of life. . . . From the

Searching for Intelligence Amid the Stars

logical, mechanical, clockwork physics of Newton, to the mysterious, profoundly perplexing universe explored by Einstein and Heisenberg, where particles pop in and out of existence and where atoms may influence each other at speeds faster than light. . . . To the sense, barely apparent even several decades ago, that these strands are not extending out in parallel fashion. They are converging.

22 | Andy Dufner, Physicist-Theologian

The scientific and technical view of the world concerns itself with the physics and engineering that drive and sustain existence. The religious believer, on the other hand, often bypasses quantifiable matters to ponder questions of purposeful design, and is less concerned with measurement than with meaning.

Yet while the scientist and the believer each have their own focus, they really are considering the same question: What's it all about?

It's pretty heady stuff to think that the forces which the scientific community identifies as gravity, electromagnetism, and the strong and weak interactions of atomic particles are of the same force that appeared in historic time in the form of a burning bush that revealed itself as "I am," that spoke to men one night in Jerusalem and said, "Peace I leave with you, peace is my gift to you."

The scientific vision and the spiritual vision give us awesome glimpses into their respective domains. But the two are

Andy Dufner, Physicist-Theologian

forced to regard each other at arm's length. We study the universe either as a spiritual or as a material order, and hold that the two orders are not congruent, concurrent manifestations of the same reality. Even though, as we understand the atom now, it is simultaneously substantial and insubstantial.

The tension between the study of the quantifiable world and the transcendental order was evident at the beginning of the computer age. Blaise Pascal was a seventeenth-century physicist and mathematician who is credited with, among other things, developing the theory of probability. An invention of his, the Pascaline, was the world's first automatic calculator, and the history of computer science is generally held to begin with that device.

Then, suddenly, at age thirty, calling his earlier interests "games and diversions," he gave up scientific pursuits to spend the remaining nine years of his life writing essays on spirituality.

Though a dramatic case, it isn't exceptional. We have come to take it for granted that the scientific and spiritual temperament can't inhabit the same mind.

Yet occasionally they do. Gottfried Wilhelm Leibniz (1646–1716) developed a calculator that was significantly advanced from the Pascaline, developed the basic theories of the calculus, and promoted the benefits of the binary system (doing it in a theological context wherein "one" represented God, "zero" the void). The man's interest ran from natural philosophy to optics to statesmanship. In his spare time, he set out to mend the then-recent rift between the Catholic and Protestant churches.

Though Pascal and Leibniz acted upon it in different ways, it is apparent that even since early in the development of computer science people in that field have had a keen, if sometimes painful, need to address their drives for mathematical precision and transcendental enlightenment.

Computer science, represented by Silicon Valley, has come a long way from the Pascaline, but we still seem to have no

Andy Dufner, Physicist-Theologian

better understanding of how to deal with the tension between these drives.

One of the few places in the country, or in the world, set up specifically to engage a conversation between the culture of science and the culture of theology is located near Silicon Valley at the Center for Theology and the Natural Sciences at the Graduate Theological Union at Berkeley.

Father Andy Dufner exudes the same calm, quiet self-assurance as Robert Noyce. After taking an undergraduate degree in physics at Gonzaga University in Spokane, Washington, he decided to enter the Jesuit order of Catholic priests. Along with his philosophical studies, he pursued his Ph.D. in theoretical physics from St. Louis University, then while studying for his licentiate degree in theology, he also did postdoctoral work in particle physics at the Stanford Linear Accelerator Center.

After several other assignments, including a guest professorship at the graduate school of physics at the University of Louvain, in Belgium, he returned to the Bay Area.

At the time we met, he was working at the Lawrence Berkeley Lab, located in the hills above the university, which he describes as a "scientific supermarket" where fundamental research is done in such fields as particle physics, energy conservation, biochemistry and medicine. Unlike the classified work on nuclear energy and weapons development done thirty miles away at Lawrence Livermore Lab, the work of the Lawrence Berkeley Lab is open and nonsecret.

We met late one afternoon at his residence, a Jesuit community located in a large house on a shaded residential street at the base of the Berkeley hills. The house was spare and simple, but certainly not monastic.

Dufner had invited me for dinner. The meal, too, was simple but not monastic, served cafeteria-style. We left the other residents, mostly middle-aged men, in the dining area and went to a large meeting room to talk alone.

Andy Dufner, Physicist-Theologian

This front room was elevated from the city blocks downhill and to the west, and out the windows at the far end of the room I could catch a glimpse of the Bay. This house seemed marvelously out of place in the vortex of social and scientific tumult that describes the Berkeley all around us.

Antagonism, or at least misunderstanding, has been a hallmark of the relationship between the scientific and spiritual pursuits. The interdenominational center with which Dufner is affiliated is set up to see what contribution science has made, or could make, to the theological vision, and, reciprocally, to see if theology might have any words of wisdom for science.

"Has science had any impact even on something as fundamental as creation theory? We find a lot of places where it has not. A four-page article in a renowned church encyclopedia purported to discuss creation of the universe by God, and had nothing to do with the physical universe as we understand it. A complete abstraction.

"I don't think there's any other cultural process that takes place that would even attempt that [divorce from scientific evidence], but theology does. I think Christian theology has always had a tough time dealing with the physical world."

But if not overtly, as now at the center, then surreptitiously through the centuries there has been an interaction.

"Looking at history, you see that the world view is put together in a large measure by scientists. And then that world view is taken over by the theological enterprise. Theologians have used concepts like time and space, matter and form, even methodology. A lot of theologians don't realize the implications of that, that their very notions of space and time, for example, are scientifically based. Or that their methodology cannot escape the same limitations discovered, scientifically, in this century for all methodology. Some people even trace back the hierarchical structure of the church as being modeled, in some sense, on the scientific world as it was known in those early days.''

Specifically, the modeling of the church hierarchy may reflect the classical, Aristotelian view of the physical universe as made up of concentric, hollow spheres, with the motion—and hence power and authority—of everything we see from earth supplied by a succession of higher and higher spheres, until finally the source of all motion is reached in the sphere of the heavens.

If it only went this far, this discussion might be of interest to a handful of scholars. But this all may have some rather immediate, here-and-now application, even to a study of Silicon Valley.

"It has always been a theological problem, a philosophical problem, as to how, if human beings are not simply matter, you can have a spirit-and-matter combination. Nobody wants to separate them off into an utter and complete duality, but usually they end up doing that. You've got to have some kind of an interaction point, or else you are constantly violating the natural laws of the universe.''

Yet while there are some theologians who study this spirit/matter interaction in the light of the new physics (which concerns itself with the wave/particle duality of all matter), whatever insights have come from that, it seems, are rarely if ever heard from the pulpit.

"Using the traditional model of spirit and matter, the question becomes: "How do I make choices, and then influence the body which is part of me? It turns out that one well-founded interpretation of the Uncertainty Principle allows for that kind of interaction to take place without the violation of the laws of nature. You get near the points in your neuronic system where these kinds of triggers occur which then cascade in ordinary material interactions, out to the point where your fingers move. You can get an interaction of spirit and matter at that point that would not violate

Andy Dufner, Physicist-Theologian

the laws of nature, and yet would allow the spiritual dimension to influence nature.''

And so it may turn out that that is where spirit and material substances interact.

''If that's the case, and that is one possible interpretation of the Uncertainty Principle, then theology has a reasonable inroad into the material world, where spirit can touch matter, whether it's the spirit of God, the spirit of angels, or human spirit. It can touch and influence matter without violating natural laws.''

And if that is true, then it may be that by means of electronic engineering—the ability we have to touch the atom in practical ways—we have been given a way to shake hands with the spiritual order, in a manner we can't even imagine now.

If theology and philosophy haven't been comfortable in the presence of natural science and technology in recent times, it is time to commence a dialogue between these two outlooks. For they are soon going to need each other's counsel.

The fundamental question we are going to face in the next few decades is this: What is it to be human? Because we are about to see our loosely defined concept of humanity—a vague notion of us as existing somewhere between apes and angels—undergo strange permutations.

As the science of robotics becomes an art, we will be able to replace many, most, all of our organic parts—perhaps even the brain—with nonorganic components. Man into machine?

At the same time, very preliminary work is going on now attempting to create electronic circuits among organic molecules. If such molecules can be grown, embedded with self-referential programming, we might conceivably grow ''living'' computers. If that could be done, placing that computer in an auto-motive body would be a relatively simple process. Machine into man?

Will we then be overwhelmed by our technology? Or have

we the wit to acquire the wisdom to know how to deal with it? Or is it possible that we were only a link in the evolutionary chain, and that we will fashion our successors?

I asked Dufner—the physicist, the priest—if upon meeting a laboratory-grown brain he would wish to baptize it. He replied without missing a beat; "I'd want to talk with it first. I'd want to find out, 'Who are you? Are you really an independent person?' "

Asked if he sees any scientific roadblocks to creating such a cybernetic unit, he said he doesn't. We'll get there gradually, he feels, but the eventual replacement of organic parts, as well as the making of lab-built "brains," is not impossible.

The precedent of a Barney Clark, living with a synthetic heart, shows the former road is not that unimaginable. It's the latter course that raises some interesting questions.

"Some philosophers would say that computers will never get there (being human-like). But when you begin talking about molecular reconstruction, about DNA research, what's the difference whether the thing grew out of a union of sperm and egg, or whether it's put together in a test tube, so to speak? Then you teach it and see how it learns, then you talk to it and find out, is there really a transcendent presence there?

"In the beginnings of those kinds of experiments, you're talking about rather crude sorts of communications. Most of us recognize a human being when we see one." And, hence, wouldn't confuse an "it" with a person.

"But if you've got to deal with it by talking to it over a terminal, or using crude voice communications, or some tonal scheme, then you've got a different ballgame. Then I think you'd better have philosophers and theologians cluster around the terminal, along with the technicians, to have their input."

That's not to say this doesn't require some concern.

"I think anywhere you deal with that kind of breakthrough in science, in which it is possible that something will get into

Andy Dufner, Physicist-Theologian

a positive feedback framework . . . where it can actually take off on its own . . . then you'd better be very careful.

"I don't think we'll be there for a long time. I know we won't. In fact, the process toward it will be very gradual, so it won't be for many years. But it's coming, I think, and it's going to be upon us. I just think we'll be a lot better prepared for it when it does come. We will have worked through the steps, we will have intermediate devices of vast capabilities before we get to something that might possibly leap off on its own."

Whenever it happens, what will it mean to us as humans?

"All we're doing then is making another indomitable human spirit emerge from the potentialities of matter in another way. We don't find computers doing that to us."

Could we not conceivably create our own masters?

"If it gets to the point where a computer can actually begin to assume control over human life processes, then we may be put in jeopardy as a race. That is not outside the realm of possibility. It's when you get to that stage that you might have something that looks utterly benevolent and stays utterly benevolent until you turn that final switch and give it control of itself. Then it says, 'Aha! Gotcha!'

"I would walk up to it very slowly, which I think is the way it will move historically, and keep an eye on it. . . . By the time we get to that point, we are going to have a much better understanding of what the implications of it are."

Which means that by the time we get to that point, we're going to have to go far beyond data processing and information processing and even knowledge processing, to wisdom processing.

23 | Conclusion

When I leave Dufner's residence, I am not in the mood to take Humphrey Go-BART from the campus back to the Berkeley BART station. The evening is far too beautiful, and it has been rare this year to have had a free evening. So I head up Strawberry Canyon, above the campus, above Berkeley, above the Bay, to the piazza in front of the Lawrence Hall of Science, perched on top of the Berkeley hills, offering one of the most splendid, sweeping views of the Bay Area. On an evening this clear, I can watch the sun set out beyond the Golden Gate Bridge, far ahead of me, and see fifty miles south to Santa Clara on my left.

As I walked up here I was aware of the evocative scent of eucalyptus on this hillside. It's a sense impression that struck me vividly on my first visit to California years ago, and one of the lures that kept drawing me back to this state from my native Minnesota.

The view before me tonight, of sunset at the Golden Gate, is one of the most representative views of the Western United

Conclusion

States. This is the end of California, the end of the Wild West, the end of the Western world. That sun is defining my evening. The next time anyone sees it will be tomorrow morning in the Orient.

But I'm not thinking about the West right now, but rather about a man who lived in New England in the nineteenth century.

Henry Adams was born in 1838, in the shadow of the Boston State House, the offspring of one of the most illustrious families in American history. In 1900, he attended an exposition in Paris and, as he later wrote in his autiobiography, *The Education of Henry Adams,* he saw there how two forces had shaped our world. The first was the Virgin, the fertile mother whose love inspired most of the great art of the Middle Ages. The other was the Dynamo, the engine that was the driving force of the Industrial Age. There was a paradox to be sure, predating Paradox Valley, that two such disparate forces had the ability to shape men's outlooks: the unselfish mother and the anonymous engine. And that's the last paradox I hope to deal with regarding this Valley.

Outside the Lawrence Hall of Science where I've come to catch the sunset, there is a single emblem that conveys both the gracious mystery of the Virgin and the impersonal power of the Dynamo. A hundred yards from where I sit now is the magnet that was the core of Ernest Lawrence's first cyclotron, a gift from the old Federal Telegraph of Palo Alto. It represents in a single form the forces Adams symbolized in the Dynamo and the Virgin. For it helped give us a glimpse of the mystery, the wonder, the sheer grandeur which underlies the material world at the subatomic level. The cyclotron represents an accomplishment of our ingenuity—an ingenuity which pales in the face of the ingenuity it revealed: the living imagination sustaining existence.

The cyclotron was also instrumental in proving the principles which provided the materials that found devastating use when an atomic bomb was dropped on Japan.

The cyclotron: a symbol of transcendental revelation, of human apocalypse.

A paradox? Rather, just the opposite. A synthesis. A device that represents a convergence of the creative and destructive in human nature.

Perhaps, too, Silicon Valley is not Paradox Valley, after all. Maybe it, too, is a synthesis, or at least points in the direction of one.

We live in a universe made up of atomic particles that are there and not "there," that are particle and wave, the ingredients in silicon and the concept behind electronic engineering, that make up existence and can destroy civilizations. Can we not, therefore, learn to understand the harmonious interconnectedness of subjects that up to now have been considered antagonistic: technology and humanism, science and art, commerce and vision, the analytic and the intuitive, the material and the spiritual?

How much longer will we pretend that technology and humanism, the study of wiring diagrams and the quest for values, are not interdependent?

Somewhere out ahead of us is a thought breakthrough waiting to be made, one that will better enable us to understand the harmony inherent in the contradictions we now stumble over in our nature. I don't know what that thought breakthrough is. It's beyond me. But so much going on now points that way.

The cyclotron is emblematic of our nature: we create and we destroy. There must be some way to sythensize those drives in us, without its spelling the end of us. The promise of that synthesis has always been out before us, though beyond most of us. Its name is wisdom. We are in a race with our own ingenuity to achieve it, or risk losing everything. If the definition of technology is the application of scientific principles to practical matters, then a definition of wisdom might be the application of appropriate values to technology.

The urge to find a synthesized overview is apparent even in

Conclusion

the natural sciences today, as physicists seek a grand unified field theory: a simple, elegant scheme to show the interconnection between gravity, electromagnetism and the strong and weak forces of the atom.

Silicon Valley may not be the first community of a new age. It may, rather, be the last community of the old age. But having, in less than forty years, led the way from data processing to information processing to knowledge processing, it may point to a time when human nature understands itself well enough to engage in large-scale wisdom processing: on the individual level, the interpersonal level, the international level. It may be that the challenge to our human identity from machine intelligence will force us to better understand who and what we are as people.

For the real significance of the technologies of Silicon Valley—of microelectronics and artificial intelligence and genetic engineering—is that we may soon become, for the first time, active architects, not passive elements, in our own evolution.

But even as we reach that level of technical agility, we develop a pressing need for more humanist learning: of history, to know what's worked and what hasn't; of communications skills, to add quality control to the message as well as the medium; and of the arts, to learn the gift of empathizing with others.

Then again, we may, using our technology, blow ourselves to kingdom come. And all the eons of effort that went into this enterprise—that astonishing cascade that brought carbon, hydrogen and oxygen atoms out of the void; that in the fullness of time saw a tiny fraction of those individual atoms merged to become breathing, sensing, wondering beings—may, at least on this planet, get short-circuited. Be a false start. Better luck elsewhere.

However much, on a day-to-day basis, the inhabitants of Silicon Valley are engaged in the business of getting and spending, however much the individual goal here is to own a

Ferrari and an estate in Los Altos Hills, the work done here is giving humankind the tools to carry out the ancient oracle's command: know thyself.

A thought breakthrough waits ahead of us, in comprehending who we are and how to proceed from here. It may come from a supercomputer, from a radio signal sent by an advanced civilization, or from another tormented, inspired loner sitting in a roadhouse in Illinois. But it will come. And when it does, it won't be the end of anything, but another ratcheting forward of the wheel of human understanding. And when it does, it may or may not occur in Silicon Valley, but it will certainly owe a debt to this community of technical scholars and inventors, artists and hackers, detectives and journalists, lawyers and scientists, financiers and philosophers. Indeed, it will.

Notes

[1]Irwin Ross, "What's New About this Boom?" *Fortune*, May 30, 1983, p. 51.

[2]Quoted in James Pooley, *Trade Secrets: How to Protect Your Ideas and Assets* (Berkeley, Calif.: 1982 [Osborne/McGraw-Hill]), p. ix.

[3]Semiconductor industry figures from Shirley Thomas, "William Shockley," in *Men of Space* (Radnor, Penn.: Chilton, 1962), p. 188.

[4]Tandem investment figures from phone conversation with Finance Department at Tandem, May 1984.

[5]Figures on the fluctuation of venture investment from Venture Economics Division of Capital Publishing Corp., of Wellesley Hills, Mass.

[6]Figures on average returns from sources at *Venture Capital Journal*, Capital Publishing Corp., Wellesley Hills, Mass. in phone conversation, May 1984.

[7]The term "avalanche of newcomers" is from John Dizard, "Do We Have Too Many Venture Capitalists?" *Fortune*, Oct. 4, 1982, p. 106.

[8]GNP of The Gambia, quoted from "World Military Expenditures and Arms Transfers, 1971–1980," U.S. Arms Control and Disarmament Agency, March 1983.

[9]*San Jose Daily Mercury*, October 12, 1891, p. 1.

[10]Of the twenty-five people profiled in this book, all were interviewed for this project with the exception of Frederick Terman. I spoke with Professor Terman by phone in late November 1982 and requested such an interview. He said his health was poor, but should he feel better after the first of the year, he would call me back to discuss our getting together. He died several weeks after that conversation, on December 19, 1982. This profile is based on material made available to me by Stanford University's News Service.

[11]Statistics on California's wartime economy are from Warren A. Beck and David A. Williams, *California: A History of the Golden State* (Garden City, N.Y.; Doubleday, 1972).

[12]An extensive account of the IBM-Hitachi case is in David Tinnin, "How IBM Stung Hitachi," *Fortune*, March 7, 1983, p. 50.

[13]The *Wall Street Journal* article on the Hitachi payment of $300 million to IBM appeared on November 9, 1983, p. 3. A news story announcing lowering of payments appeared in the *San Francisco Chronicle* on June 12, 1984.

Notes

[14]Herbert Swartz, "High-Tech Law Requires an In-House Law Department," in *Electronic Business*, May 1, 1983, p. 76.

[15]A brief overview of the Pooley/IBM case is in Kathleen Sullivan, "IBM Launches Secrets' Suit," *Management Information Systems Week*, January 4, 1984, p. 4.

[16]The news that a suspect had been arrested in the case of the Skyline Boulevard murder appeared in the *San Jose Mercury News*, June 21, 1984, page 1B.

[17]An account of the Seagate case can be found in Katherine Hafner, "Seagate's Corporate Scientist Unmasked," *Computerworld*, August 8, 1983, p. 13.

[18]The incident of the angry neighbor is recounted in Ivan Sharpe, "Yuri's People: Life with the KGB in Pacific Heights," *San Francisco Sunday Examiner and Chronicle*, August 21, 1983, p. 1.

[19]Stewart Brand, *II Cybernetic Frontiers* (New York: Random House, 1974).

[20]The IBM explosion was reported in the *San Francisco Chronicle* on June 14, 1982, p. 1, and June 15, p. 5.

[21]David Walworth, "Chemical Warfare," *Bulletin of the Santa Clara County Medical Society*, April 1982, p. 4., gives details of the Fairchild leak.

[22]The figures come from San Jose mayor Tom McEnery, quoted in Haynes Johnson, et al., "America Today" *Washington Post*, September 18, 1983, p. 1A.

[23]Source for defense expenditures is *Prime Contracts Awarded*

over $25,000 by State, County, Contractor and Place, Fiscal Year 1983, available from the Directorate for Information Operations and Reports, Pentagon, Washington, D.C. 20301. Employee figures are from *The Answer Book for Santa Clara County, 1984 Edition*, published by the *San Jose Mercury News*, p. 50.

[24]Melman's figures are quoted in "Looting the Means of Production," which appeared in the Mid-Peninsula Conversion Project newsletter, *Plowshare Press*, November-December 1982, p. 8.

[25]One source Yudken cited for these figures on research and development and comparative gross domestic product is Robert DeGrasse et al., *Military Expansion, Economic Decline* (New York: Council of Economic Priorities, 1983).

Bibliography

Adams, Henry. *The Education of Henry Adams*. Boston: Houghton Mifflin, 1918.

Beck, Warren, and David A. Williams. *California: A History of the Golden State*. Garden City, N.Y.: Doubleday, 1972.

Boly, William. "The Gene Merchants." *California Magazine*, September, 1982.

Brand, Stewart. *II Cybernetic Frontiers*. New York: Random House, 1974.

Bylinsky, Gene. *The Innovation Millionaires: How They Succeed*. New York: Scribner, 1976.

Cole, Tom. *A Short History of San Francisco*. San Francisco: Lexikos, 1981.

Davis, Henry B. O. *Electrical and Electronic Technologies: A Chronology of Events and Inventors to 1900*. Metuchen, N.J.: Scarecrow, 1981.

Dillon, Richard. *Humbugs and Heroes: A Gallery of California Pioneers*. Garden City, N.Y.: Doubleday, 1970.

Bibliography

Durrenberger, R.W. *The Geography of California in Essays and Readings*. California: Brewster, 1959.

Ellul, Jacques. *The Technological Society*. Tr. by John Wilkinson. New York: Vintage, 1967.

Gardner, W. David. "The Independent Inventor." *Datamation*, September 1982.

Gillispie, Charles Coulston, ed. *Dictionary of Scientific Biography*. New York: Scribner, 1976.

Gomes, Lee. "Secrets of the Software Pirates." *Esquire*, June 1982.

Hale, John. *The Renaissance*. New York: Time Inc., 1965.

Hofstadter, Douglas R. *Gödel, Escher, Bach: An Eternal Golden Braid*. New York: Vintage, 1980.

Hoefler, Don C. *Silicon Valley Genealogy*. A chart of the genesis of semiconductor companies. Mountain View, Calif.: Semiconductor Equipment and Materials Institute, Inc., 1982.

————. "Silicon Valley, USA." 3-part series. *Electronic News*, January 11, 18, and 25, 1971.

Johnson, Paul C. *A Pictorial History of California*. New York: Crown/Bonanza, 1970.

Kim, Scott. *Inversions: A Catalog of Calligraphic Cartwheels*. Peterborough, N.H.: Byte Books, 1981.

Kinnard, Lawrence. *History of Greater San Francisco Bay Region*. 2 vols. New York and West Palm Beach: Lewis Historical Publishing, 1966.

Kotkin, Joel, and Paul Grabowicz. *California, Inc.* New York: Rawson, Wade, 1982.

Larner, John. *Culture and Society in Italy, 1290-1420*. New York: Scribner, 1971.

Miskimin, Harry A. *The Economy of Early Renaissance Europe, 1300–1460*. Englewood Cliffs, N.J.: Prentice-Hall, 1969.

Morgan, Jane. *Electronics in the West: The First Fifty Years*. Palo Alto, Calif.: National Press, 1967.

Morrison, Philip, John Billingham, and John Wolfe, eds. *The

Bibliography

Search for Extraterrestrial Intelligence: SETI. Washington, D.C.: NASA Scientific and Technical Information Office, 1977.

Newman, James R., ed. *The Harper Encyclopedia of Science*. New York: Harper & Row, 1967.

Nunz, Gregory J. *Electronics in Our World: A Survey*. Englewood Cliffs, N.J.: Prentice-Hall, 1972.

Overhage, Carl F. J., ed. *The Age of Electronics*. New York: McGraw-Hill, 1962.

Parker, Donn B. *Fighting Computer Crime*. New York: Scribner, 1983.

Pirenne, Henri. *Economic and Social History of Medieval Europe*. Tr. by I. E. Clegg. New York: Harcourt, Brace & World.

Pooley, James. *Trade Secrets: How to Protect Your Ideas and Assets*. Berkeley, Calif.: Osborne/McGraw-Hill, 1982.

Pratt, Stanley, ed. *Guide to Venture Capital Sources*, 5th ed. Wellesley Hills, Mass.: Capital Publishing, 1981.

Rambo, F. Ralph. *Almost Forgotten: Pen and Inklings of the Old Santa Clara Valley*. 1964.

Rosenbaum, Ron. "Secrets of the Little Blue Box." *Esquire*, October, 1971.

Rybczynski, Witold. *Taming the Tiger: The Struggle to Control Technology*. New York: Viking, 1983.

Servan-Schreiber, Jean-Jacques. *The World Challenge*. New York: Simon & Schuster, 1980.

Shockley, William. "The Path to the Conception of the Junction Transistor." IEEE Electron Devices Society, 1976.

Snow, C. P. "The Two Cultures." *New Statesman and Nation*, October 6, 1956.

Stavrianos, L. S. *The World Since 1500: A Global History*. Englewood Cliffs, N.J.: Prentice-Hall, 1966.

Thomas, Shirley. "William Shockley." In *Men of Space: Profiles of the Leaders in Space Research, Development and Exploration*. Radnor, Pa.: Chilton, 1962.

Wolf, Fred Alan. *Taking the Quantum Leap*. San Francisco: Harper & Row, 1981.

Zientara, Marguerite. *The History of Computing*. Framingham, Mass.: CW Communications, 1981.

Zukav, Gary. *The Dancing Wu Li Masters: An Overview of the New Physics*. New York: Morrow, 1979.

Acknowledgments

A great many people contributed to this book, more than I can name here, but I do wish to offer grateful acknowledgment for the generous help which the following people provided.

First, my thanks to the people profiled here, for sharing with me their time, their thoughts and their enthusiasm.

Many others provided extensive background information, or pointed me in the right direction: Clyde Arbuckle, local historian; Austen Warburton, authority on Costanoan life; Franklin ''Pitch'' Johnson, for insights into the transformation that occurred in this valley; Thorn Mayes, for information on Lee de Forest; Bob Beyers of Stanford University's News Service for providing excellent background on Fred Terman; Bill Moulton of Waveform Corp., for his thoughts on the role of the computer in the arts; Kathy Lusk of the Museum of Electronics at Foothill College; Ed Clark of the National Weather Service; Howard Stevenson of the Harvard Business School; Donn B. Parker of SRI International; Jane

Acknowledgments

Koloski Morris of *Venture Capital Journal*; Tom Mysiewicz, editor of *Bio Engineering News*; Ken McKenzie of Dataquest; Dr. Patrick Williams of the Center for Organization and Management Development at San Jose State; Ricka Pirani of the California Employment Development Department; Judith Singer of the California Department of Economic and Business Development; Stan Baker of *Electronic Engineering Times*; Rory O'Connor of *InfoWorld*; Dr. Roger Vass; Jose Ramos; Judith Goldhaber and others at Lawrence Berkeley Laboratory; Marie Getchel; Madelen Schneider; Jim Jarrett; Dr. Charles Versaggi; Matthew McIntosh; Joan Green; Marcia Todd; Roy Verley; George Boardman; and representatives of many of the companies mentioned herein for providing background information.

I'd particularly like to thank Peter Carter and Edna Markham of Carter, Callahan Advertising and Public Relations in San Jose, for generously allowing me to rearrange my schedule there while I worked on this project. Too, my thanks to my co-workers, friends and clients at Carter, Callahan who were so supportive, and so understanding of my eccentric hours.

I appreciate, too, the excellent work done by the research staffs at the San Jose, Sunnyvale and Contra Costa County public libraries, and the Bancroft Library at UC-Berkeley. Their attention to detail is surpassed only by their promptness and professionalism. Maria Pursley, Glenna Goulet and Mary Hennessey were each kept busy typing dozens of hours of tape transcripts.

My thanks to John Dodds, Daniel Frank and Mitchell Rose at New American Library; and especially to Maureen Baron at NAL. Other friends who, directly or indirectly, made this possible include Brian Richard Boylan, Ben Marsh, Al Kane, Ed Addeo, Jim Forbes and, most particularly, Carole Abel.

I'd also like to acknowledge Robert M. Pirsig, who showed

Acknowledgments

a generation of writers that matters technical can be explained not only lucidly, but elegantly.

And finally, to my mother, my brother and my late father; and to my wife, Mary, and our sons, Christopher and Joseph, I dedicate this book, in appreciation and with love.

Tom Mahon
San Francisco, California

Recommended MENTOR Books